OIL AND THE BRITISH ECONOMY

OIL and the British Economy

FRED ATKINSON and STEPHEN HALL

CROOM HELM
London & Canberra

ST. MARTIN'S PRESS
New York

© 1983 Sir F. Atkinson and S. Hall
Croom Helm Ltd, Provident House, Burrell Row,
Beckenham, Kent BR3 1AT
Croom Helm Australia, PO Box 391,
Manuka, ACT 2603, Australia

British Library Cataloguing in Publication Data

Atkinson, Fred
    Oil and the British economy.
    1. Petroleum industry and trade – Great Britain
    2. Great Britain – Economic conditions
    I. Title II. Hall, Stephen
    338.2'7282'0941   HD9571.5
    ISBN 0-7099-0528-9

Library of Congress Cataloging in Publication Data

Atkinson, Fred, 1919-
    Oil and the British economy.

    1. Petroleum industry and trade – Government policy –
Great Britain. 2. Gas industry – Government policy –
Great Britain. 3. Great Britain – Economic policy – 1945-
I. Hall, Stephen, 1953-   II. Title.
HD9571.6.A84   1983   338.2'728'0941   83-11044
ISBN 0-312-58295-1

CONTENTS

Our thanks are due to the National Institute of Economic and Social Research (NIESR) who kindly allowed us to use their model to perform the simulations presented in Chapters 7 and 8. It should be stressed that the Institute bears no responsibility for any part of this book.

INTRODUCTION

The discovery of oil and gas in the North Sea and the rest of the continental shelf was greeted as a stroke of wonderfully good luck for this country. It was widely expected that the development of these finds would bring in a period of prosperity for the British economy. In fact, the early years of the 1980s, when oil production has risen to the level of self-sufficiency and beyond, have been years of recession and unemployment. In this book we investigate the connection between oil and the rest of the economy in order to explain why these hopes were disappointed. We analyse the economic policy of the government and discuss alternative ways of managing the economy in order to get maximum benefit from our oil and gas wealth. The book is in three parts. The first part deals with oil and gas simply as natural resources. The second part examines the effect of oil and gas development upon the industrial sector, both in this country and in certain others. The third part is concerned with macroeconomic effects and policies.

Part I

To begin with we look back in Chapter 1 to see what has happened up till now in respect of oil and gas. The gas finds came first. The Gas Council, predecessor of the present British Gas Corporation, was made monopoly buyer and settled a low price with the oil companies which were bringing the gas ashore. The effects of the gas on the economy are difficult to pin down. The gas is considerably cheaper than the equivalent in coal would have been; the profits of the gas industry rose sharply and helped to pay for the conversion to natural gas; the balance of payments gained from the reduction of oil imports and the domestic consumer enjoyed the main benefit by getting a cheap form of energy.

Oil production began in 1976 and built up to 89.4 million tonnes a year by 1981, considerably exceeding home consumption. In the middle years of the 1970s the UK had suffered very severe balance of payments problems as the price of oil shot up and added greatly to the cost of imports. This burden was totally removed by the increase in

7

North Sea production of oil. Of course, oil from the North Sea costs a good deal to produce, especially in terms of capital expenditure. It was dearer than the oil imported into Britain before the big price increase of 1973 but if it had not been for North Sea oil the UK would have had to meet an enormously increased import bill as other countries have had to do. A vast effort of adjustment would have been needed to pay this bill. This would have meant much worse terms of trade and possibly a lower level of activity as well.

While the effects on the balance of payments and national income have come through in a big way, the benefit in the form of increased tax revenue has been delayed but it is now rising rapidly and in 1981/2 it is estimated at some £6.4 billion or 6.5 per cent of total tax revenue.

In Chapter 2 we examine the prospects for oil and gas, and the first question to consider is the size of reserves on the UK continental shelf. According to the official estimates oil reserves may be estimated at between 2,100 and 4,300 million tonnes. At 1980 rates of consumption these stocks would last from 26 to 53 years. This is probably a conservative estimate and the true figure for reserves could be near the top of the official range. Hence policy should be based on the assumption that the Continental Shelf oil deposits are large and will last well into the twenty-first century.

As regards gas, the official estimate for remaining reserves is of a range of some 1125-1875 million tonnes of oil equivalent, equal to 27 to 45 years' consumption at the 1980 rate. Again, a figure near the top of the range is the most likely. There is also great uncertainty about future rates of oil production. Even over the next five years or so when production will come from established fields there are considerable variations in estimates from different expert sources. Most experts envisage production of the order of 100 million tonnes a year or more.

The taxation system has become extremely complex and future tax revenues even in the short term are very difficult to predict. Assumptions have to be made about future world oil prices, exchange rates, production levels, capital expenditure and inflation. The Treasury's most recent estimate of government oil revenue in 1983/4 is about £6 billion. Other estimates are far higher, ranging from £10 billion upwards. As in the case of the forecasts of oil production most outside experts consider the official estimates to be very much on the low side.

So much for the prospects for output and revenues over the next few years. A big issue now arises: how should these massive resources be managed? Is it wise to use up the riches of the North Sea in current consumption or should the proceeds be invested so as to provide a continuing income? Should depletion be deliberately delayed so as to keep oil and gas for our longer term future? Chapter 3 is devoted to these questions.

It seems reasonable that the benefits of oil and gas should be shared between present and future generations, since these resources are irreplaceable. To the extent that the resource is saved, it can be invested so as to yield a return. One possible criterion for the amount to be consumed currently would be the possible yield on the value of the existing stock of oil. If this yield were for instance 1 per cent then £2,500 million a year could be consumed indefinitely without reducing the real value of the asset.

The question of how the oil and gas resources should be invested is difficult to answer with confidence. To the extent that the price of oil and gas is likely to rise more than the price of other assets and to exceed the general rate of inflation, investment in oil and gas would produce a yield in real terms. In recent years this yield has been spectacular, but the future rate of appreciation is likely to be much less. Equally the probable future yield from alternative investments is extremely hard to assess. We estimate it conservatively at between 1 and 2 per cent. Oil may do better than this but compared with many alternative investments it is very hard to realise if needed.

One of the deciding factors in making the investment decision is the security aspect of North Sea oil. In view of the world's heavy dependence on Middle East oil there is a case for preferring oil in the ground to alternative investments. This brings us to a view about depletion policy. At present production is forecast at some hundred million tonnes a year, rather more than domestic consumption. No major interference with this plan seems to be justified. A drastic reduction of output would create serious adjustment problems for the economy as the balance of payments deteriorated and would do great damage to the exploration effort. Nor would an acceleration of output be wise, given the difficulty of ensuring that the proceeds are invested in alternative ways rather than consumed and given the advantage of holding oil for reasons of security. The soundest depletion policy would seem to be one of preventing production from rising much above the self sufficiency level while encouraging exploration and production over the longer term. The government's statements on the subject suggest that this is their aim.

Part II

Chapter 4 deals with the effect of the discovery and exploitation of oil on the industrial sector of the economy. There has been much debate about this in recent years: doubt has even been thrown on the ultimate value of the oil resource in so far as it might have been responsible for the decline in manufacturing industry.

We first considered this issue in simplified terms in order to clarify the basic economics. We show that if the existing balance of trade is disturbed by the growth of a new export or import-saving industry the process of restoring the balance would involve the weakening of the balance of trade in the existing export and import lines. That is the oil, or whatever the new source of export earnings may be, will necessarily displace some of the traditional export products and lead to the import of some things that were previously produced at home. If, as is the case in this country, the traditional exports consist mainly of manufactured goods, then the manufacturing sector will contract somewhat.

But it is a far cry from this simplified analysis to the case of British oil in the 1970s. The rapid growth of oil production in this country was not superimposed on a state of balance of payments equilibrium, but on a state of substantial deficit, partly caused by the great increase in the price of imported oil. Thus British home produced oil helped to get rid of this deficit rather than to cause a surplus in the 1970s: there was no need for existing exports to be displaced. In any case the introduction of home produced oil took place not suddenly but over a period of several years in which there would normally be some growth of the economy including growth of manufacturing. Over a period of several years the result might be a relative decline in manufacturing but not a fall in the absolute level.

Later on, between 1979 and 1981, manufacturing production did in fact fall sharply in Britain, but this was mainly the result of restrictive economic policies and the consequent high exchange rate. Oil production played some part in pushing up the exchange rate. The recession should also be seen against a background of a long-term trend of decline in the share of manufacturing in total output which was apparent well before the arrival of North Sea oil.

In Chapter 5 we make an excursion abroad to see how other countries have been affected by the growth of a natural resource.

Norway is perhaps the closest analogy to the UK. Oil production there began four years earlier than in the UK and represents a higher proportion of GDP and exports. The exchange rate had an upward trend from 1970 to 1977, but there was no decline in the tradeable sector of production. Full employment was maintained, largely as a result of deliberate government policy in the form of loans and subsidies to industry. It is questionable how viable Norwegian industry would be if the subsidies had eventually to be withdrawn. Certainly productivity had flagged.

The Netherlands authorities took a rather different line with their natural gas discoveries. While attempting to maintain a fairly high level of demand overall they did not intervene directly in the industrial sector. Gas exports on a massive scale led to a considerable appreciation of the exchange rate during the 1970s, displacing some traditional exports. Nevertheless, manufacturing output increased substantially over the decade, keeping pace with that of Germany, for example. This has been accomplished with a smaller labour force thanks to a good increase in productivity and big changes in the industrial structure. At the same time large surpluses were earned on the balance of payments: these were invested abroad, so preserving some of the benefits of the natural gas endowment.

Three countries examined more cursorily in this chapter are Australia, Japan and Venezuela.

Part III

The remainder of the book deals with the management of the economy in recent years and the prospects ahead. Chapter 6 examines the link between oil and the economy at large, and then the theories of how the economy works. Chapter 7 traces government policy and its results up to the present and includes forecasts of macroeconomic development up to 1986. Chapter 8 looks at alternative strategies in which the benefits of oil and gas are used to expand the economy.

Chapter 6 begins with a look at the two main ways in which the growth of oil production affects the economy. One way is through the exchange rate; the other is through tax receipts and fiscal policy.

The increasing production of North Sea oil must have contributed to the extraordinary rise in the exchange rate in recent years because the oil greatly strengthened the current account of the balance of payments, but over the same period UK interest rates have been high and the UK has attracted a good deal of short-term capital. It is impossible to determine exactly in what proportions these two different factors shared the responsibility for the rise in the exchange rate.

From 1977 to 1980 the exchange rate rose by 25 per cent while British costs and prices rose more than those in major competitor countries. This loss of competitiveness had a devastating effect on British industry, slashing its profitability and curtailing its production. The rise in prices and in the exchange rate may have caused up to one-third of the recession which brought the GDP in 1981 to 5 per cent below that of 1979. Equally the rise in the exchange rate made a major contribution to the moderation of inflation. Without it the price level in 1981 could well have been some 10 per cent higher than it actually was.

The second channel through which North Sea oil is affecting the economy is through the budget. Receipts from the complex series of taxes on oil production have risen sharply in recent years and are projected to reach high levels over the next few years, of the order of £10 billion, or about 10 per cent of total tax receipts.

The way in which these massive revenues are used is a key factor in the management of this resource and of the economy as a whole. At one extreme the revenues can be used to finance new expenditures, whether on investment projects or on the improvement of public services; or to reduce taxes. At the other extreme the revenues from the North Sea can be used to reduce the budget deficit and what is known as the public sector borrowing requirement. The present government is intent on the latter course.

The rationale of this policy can only be understood if the government's general attitude to economic policy is explained. This explanation, and a critique of monetarist arguments, occupies the remainder of the chapter. The government's overriding aim of reducing the public sector deficit and the borrowing requirement is shown to be a departure from the pure monetarist doctrine and one which has depressing effects on the level of activity and employment.

Chapter 7 follows this discussion of economic theory with an analysis of the present government's economic measures, largely guided by its medium-term financial strategy. From the beginning the government made it clear that the way it proposed to reduce inflation was by reducing the rate of growth of the money supply. In support of this aim the government considered it essential to reduce the budget deficit year by year. Thus in successive annual budgets the Chancellor of the Exchequer sought either to reduce public expenditure or to increase taxation or both. The cumulative tightening of the fiscal stance over the four years was very severe. The result of this and of the high interest rates needed to limit the growth of the money supply as well as of the strong exchange rate, was the most severe recession and the highest level of unemployment since the 1930s.

Inflation was gradually brought down but was still around 6 to 7 per cent by the end of 1982. Thus, up to that point, the policies seemed to have failed, in that the gains were slight and the costs very great. However, the policy was essentially medium term. In order to view it in a longer perspective projections of the likely development of the economy up to 1986 are presented. These suggest that the recovery from the recession will be slow and weak and that unemployment will continue to rise. Inflation should remain in single figures and with the help of North Sea revenues the PSBR will be sharply reduced. Indeed it is the policy of reducing the PSBR which limits the extent of the recovery from the recession.

Chapter 8 is devoted to the description of alternative economic policies in which the benefits of North Sea oil are used to assist in the growth of the economy, both in the short term through more employment and in the long term through more investment in the country's capital stock which should increase productivity. The general lines of alternative, more expansionary policies, have been set out in many publications. The contribution of this chapter is to present a number of simulations of the effects of various types of stimulants to demand, using the National Institute econometric model. The general conclusion which emerges from these simulations is that the particular method of reflation is not so important as the amount. The amount of reflation would need to be very large to bring unemployment down substantially from the very high level that has now been reached.

Part I

OIL AND GAS

1 A BRIEF RETROSPECT OF OIL AND GAS IN THE BRITISH ECONOMY

The history of oil and gas exploitation in this country goes back much further than is commonly recognised. The geological conditions in which oil and gas arise are similar. A number of small gas finds were made after World War I culminating in a significant discovery in 1937 at Eskdale in Yorkshire, which established the possibility of further finds of both oil and gas. This possibility was quickly confirmed by the discovery of a number of small gas fields near Edinburgh and finally, shortly before the outbreak of war, a small field was found at Eakring, near Nottingham. The prospect which was already beginning to emerge at the beginning of World War II was, therefore, one of a high probability of widespread small oil and gas deposits. There might also be large deposits, but the state of current geological knowledge and the difficulty of drilling on many of the best sites in a built up country such as the United Kingdom made this highly speculative. Nonetheless the government was sufficiently aware of this possibility to pass the Petroleum (Production) Act of 1934. This act vested the ownership of any oil and gas discovered within Great Britain with the Crown. At the time this referred only to deposits within the three-mile limit.

The prewar pattern continued almost unaltered into the late 1940s and throughout the 1950s. A series of further small finds was made in the Midlands, and in the 1950s slightly larger finds were made at Gainsborough in Lincolnshire and Kimmeridge in Dorset. In fact by the early 1960s, total onshore oil production was running at approximately 1,500 barrels a day (between 1 and 2 per cent of total UK oil requirements). The pattern of widespread small hydrocarbon finds seemed, therefore, to be well established. There was still, however, considerable optimism about the likely extent of future discoveries of both gas and oil. This is demonstrated in part by the annexation of Rockall in 1955, a move which greatly increased the total area of the continental shelf which could subsequently be claimed by Britain. The Geneva Convention of 1958 laid down the basic guidelines which were to govern the determination of international boundaries on the sea bed of the continental shelf. The basic decision was that 'Boundaries shall be determined by application of the principle of equidistance from the nearest points of the baselines from which the breadth of the territorial sea of each state is measured.' This Convention was subsequently ratified by most European countries, including the UK, in 1964.

The Development of Gas

The pattern of a large number of small gas and oil finds was broken in 1959 when a large discovery of gas was made at Groningen in Holland by a Shell-Esso consortium. This discovery positively established the existence of major deposits of hydrocarbon fuels on the European continental shelf. This had great importance for possible UK resources as Britain is part of the same geological formation as Holland. The European governments took some time to adjust to the new conditions and possibilities which Groningen presented. But in 1964 the Geneva Convention was ratified and the first set of UK offshore exploration licences was issued. Virtually the whole of the area which lay in un-contested waters was made available. The more distant areas had to wait until 1965 when boundary agreements were reached with Norway, Germany, Denmark and the Netherlands. Agreement with the French was not reached until after the results of a court arbitration in 1977-8.

The rights to prospect for gas are allocated in exactly the same way as the rights for oil exploration. Licences for exploration and produc-tion are granted under the 1934 Petroleum (Production) Act and the 1964 Continental Shelf Act (amended by a number of subsequent sets of regulations). The licences are of two types: exploration licences, which only allow preliminary geological survey work, and production licences, which are necessary for any large scale drilling activity. The granting of a production licence does not therefore mean that oil or gas has been positively discovered, but only that deep exploration drilling may take place. The licensees who carried out the early gas exploration were, by and large, established oil companies. This is illustrated by the companies which have made the major gas finds: BP (West Sole), Shell/Esso (Lemon Bank), BGC/Amoco (Indefatigable), etc.

Only one well was actually drilled in 1964 and this met with no success. In 1965 a further ten wells were drilled and the first signifi-cant UK gas find was made (the West Sole field operated by BP Petrol-eum Development Ltd.). In February 1966 British Petroleum signed a contract to sell the gas from the West Sole field to the Gas Council for a price of 5d per hundred cubic feet. This was a very favourable price from BP's point of view, as it was very close to being comparable with world oil prices.

As the possible size of the total gas reserves began to be apparent there were many suggestions to alter the tax structure so that much of the excess profits could be removed from the companies undertaking the exploration. One proposition, for example, was to set up a joint consortium of the National Coal Board, British Petroleum and the Gas Council under the name of the State National Hydrocarbon Council. An interdepartmental government committee was set up to examine these proposals and make recommendations for any desirable changes. No formal changes were made to the tax structure or the method of organising the gas industry; the existing 1934 legislation was deemed sufficient when coupled with the 1958 Geneva Convention. The main policy objective was taken to be the establishment of the lowest possible gas prices to the consumer. In pursuance of this objective the Gas Council drastically cut its subsequent offers to the drilling corporations. In 1967 the best offer made by the Gas Council for the purchase of natural gas was 1.8d per hundred cubic feet.

The Gas Corporation was to all intents and purposes in the position of a pure monopsonist. The gas producing companies had no alternative buyer and the only alternative to selling to the Gas Board was not to sell at all. There was therefore no need to introduce a complex tax structure to remove the economic rent from the companies as this objective could be achieved much more effectively and easily by fixing the purchase price instead. This is particularly true as it was quite possible to vary the price between fields, even though the gas was exactly the same, so that fields such as Indefatigable, which were further from the shore than most other fields, were paid a higher price.

The arrival of natural gas presented a number of options. The gas could be depleted at a fairly low rate and supplied only to specially selected large commercial users, either for energy or chemical uses. This approach would entail fairly low levels of capital investment and allow a high price to be charged for the gas. A second approach would be to use the gas in large-scale gas power stations. This would also involve comparatively little capital investment but it would not make use of the special qualities which gas possessed and it would not be a very efficient use of a fuel which can effectively be used on site. The third option was to engage in the massive levels of investment necessary to convert all the gas appliances in the country to use natural gas (which had almost twice the thermal value of traditional gas). This option could only be justified if the consumption of gas were to rise sharply in order to repay the high capital investment. This option would therefore entail high depletion, which would be helpful to the balance of payments, and a low price for the gas in order to stimulate the new demand. This final option was the one chosen. The low gas price could be maintained by virtue of the Gas Corporation's monopsony. This option did, however, mean that the main advantages of natural gas were passed directly on to the consumer. The nation also benefited from the improvement in the balance of payments resulting from the benefits of gas for oil. Figure 1.1 illustrates the rapid growth which occurred in the natural gas sector between the middle of 1968

Figure 1.1: Inland Consumption of Primary Fuel for Energy Use

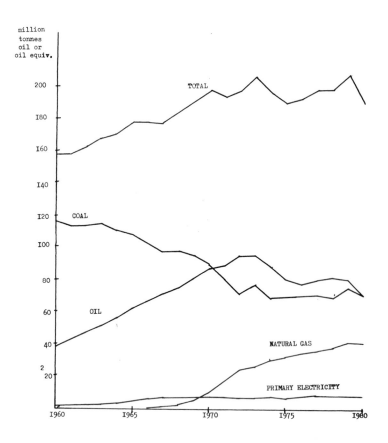

and 1972. Although the use of oil did not actually decline during this period, there was a clear reduction in the growth of the oil sector from 1970 onwards, and this represented a positive gain to the balance of payments.

Gas had, of course, been widely used as fuel prior to the discovery and development of natural gas. Figure 1.2 shows the drastic change which occurred in the gas sector between 1960 and 1975. Prior to the arrival of natural gas the traditional source of gas supplies had been coal gas. Its share was fairly stable between World War II and 1960. In the early 1960s a change began to occur in the types of gas being used. Gas produced from oil began rapidly to increase its share in total production, as did imported gas. In the late 1960s gas produced from oil was actually the major single source. As North Sea gas came on stream during the last half of the period there was both a rapid decline in the use of manufactured gas and a large total increase in gas consumption. The advent of major quantities of natural gas therefore revolutionised the gas industry in two major directions. Firstly, it was transformed from a producing industry (i.e. one which was largely concerned with manufacturing gas) into a distributive industry which

Figure 1.2: The Build-up of Natural Gas Production

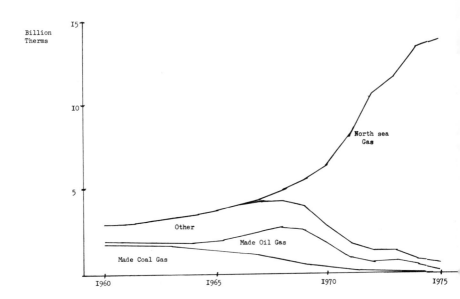

was concerned with the transportation and marketing of gas. Secondly, it virtually quadrupled the actual volume of gas being consumed, over a five-year period, thereby turning the gas industry into a major growth sector.

When trying to provide numerical estimates of the contribution of natural gas to either the nation's gross domestic product or to its balance of payments, the choice of the appropriate price used is a crucial and very difficult one.

There are basically three possible prices which might be used; the first is simply the average price actually paid for North Sea gas. The problem here is that because of the extreme monopsony power of the Gas Board this price is a wholly artificial concept which bears no relationship to the economic worth of the gas. As an example of this, in 1979 the average price paid for Southern North Sea gas was around £9/tonne oil equivalent, while Northern North Sea gas was £58/tonne oil equivalent. The Gas Board sets these prices so that the costs of extraction are covered with only a reasonable profit margin. Using this price to value gas therefore produces almost no net addition to GDP and it is quite irrelevant to calculating the balance of payments effect. The second possible price which might be used is the price of imported gas. This price does represent a free market price for gas although the international gas market has never been as large or competitive as the oil market. This means that if the UK were to try and sell all North Sea gas on the world market, there would almost certainly be some changes in world gas prices. Nonetheless, the price of imported gas is a good guide to the level of price which North Sea gas might fetch. The third possible price which might be used is the price of an equivalent quantity of energy imports in terms of oil. The main argument for this is that if North Sea gas had not existed then we would certainly not have imported such large quantities of gas but rather an equivalent quantity of energy in the form of oil.

If the object of the exercise is to assess the effect of North Sea gas on the economy then the average price actually paid for gas is clearly not the appropriate price to use. This leaves the choice essentially between the price of imported oil or the price of imported gas. These two alternatives give very different answers when the assessment of GDP contributions and balance of payment effects are made. In the first seven years of Table 1.1 the imported gas price is much higher than the imported oil price, and so using the gas price would attribute a much larger effect to natural gas. After this period, however, the situation is reversed with oil prices being higher. The choice of the

Table 1.1: North Sea Gas Production and Value

	67	68	69	70	71	72	73	74	75	76	77	78	79	80
Production Million toe	.39	1.75	4.39	9.62	15.62	22.61	24.62	29.74	30.98	32.83	34.44	32.9	33.52	31.87
Average price of North Sea gas. £toe	10.9	8.6	6.2	5.4	5.3	5.1	5.7	6.6	7.6	8.6	12.0	18.7	22.8	23.6
Value of gas £m	4.3	15.1	27.2	51.9	82.8	115.3	140.3	196.3	235.4	282.3	413.3	615.2	764.3	752.6
Recorded contribution of oil and gas to GDP £m	—	—	—	24	18	57	40	11	−7	596	2079	2771	5680	8762
Average price of oil imports £/t	6.55	7.56	7.08	6.81	8.66	8.76	11.44	33.62	38.67	51.11	57.91	53.56	63.42	95.84
Average price of gas imports £/toe	12.0	14.2	13.8	13.9	13.7	13.2	13.2	13.7	18.2	22.1	35.1	47.6	48.1	57.5
Value of North Sea gas at imported gas price £m	4.7	24.9	60.6	133.7	214.0	298.5	324.0	407.4	563.8	725.5	1208.8	1566.0	1612.3	1832.5
Value of gas at imported oil price £m	2.6	13.2	31.1	65.5	135.3	198.1	281.7	999.9	1198.0	1678.0	1994.4	1762.1	2125.8	3054.4
Adjusted GDP oil & gas contribution £m	—	—	—	106	149.2	240.2	223.7	222.1	321.4	1039.2	2874.5	3721.8	6528.0	9841.9

Source: Digest of UK Energy Statistics; authors' calculations; Development of the Oil and Gas Resources of the United Kingdom, 1982.

appropriate price rests mainly on the question being asked. If we are simply asking what is the true value of the gas then the imported gas price is the most appropriate price. But if we want to make a comparison of how the economy would have developed without gas, then the price of a close substitute such as oil may be more appropriate.

Because of the artificially low gas price it is misleading to assess either the GDP or balance of payments contributions on the basis of simple gas revenues. The GDP contribution can be most easily corrected of the two; despite the lack of a world gas market the UK does import considerable quantities of gas which, of course, has a clearly defined price. It is reasonable therefore to use this price in calculating the actual value of North Sea gas production. The sixth row in Table 1.1 shows the price of imported gas. This series bears no relation to the price of North Sea gas. The cost of imported gas is considerably higher than the price paid for North Sea gas throughout the whole period, this emphasises the artificial nature of North Sea prices. Using this price to adjust the GDP figures from Table 1.1, row 4, a revised estimate of the GDP contribution of oil and gas is presented in the last row. These figures are considerably higher with the difference approaching one billion pounds in 1978. The direct GDP contribution of North Sea gas is therefore seen to be quite considerable.

It would be tempting to use this same price series to assess the balance of payments contribution of gas as well. This procedure, however, has a conceptual flaw which makes an accurate assessment of the balance of payments position very hard to achieve. What we are trying to measure is how the balance of payments changes because of North Sea gas. It would only be correct to estimate this by the total value of the gas at imported gas prices if it is assumed that the same level of gas would have been consumed by importing gas. This, however, is not a valid assumption because the higher price of imported gas would have caused a large reduction in demand. It is not, however, sufficient to estimate the demand function for gas and use this to calculate how much gas would have been used because the change in gas prices would have been so large as to have affected the demand for other energy sources as well. In fact, as Table 1.1 shows, the price of imported oil before 1974 was considerably less than that of gas. So if North Sea gas had not been exploited it is unlikely that this extra need for energy would have been met by imported gas: it would have been met by imported oil. Table 1.1 shows the value of North Sea gas calculated at both the imported gas and the imported oil price. If there had been no North Sea gas then undoubtedly the low oil price between 1967-73 would have prevented the growth of the gas industry and the balance of payments effect may be most closely assessed by the value of the gas at oil prices. During the period 1967-73 this value is much lower than the corresponding gas price figure so the balance of payments effects can be seen to be relatively small. However, if this train of logic is followed then when the oil price rose in 1973/4 there would have been no possibility of switching to gas consumption, so oil imports would have had to be continued. The balance of payments contribution of natural gas then becomes very considerable during the period 1974-80.

The Development of the Oil Sector

The discovery of oil in large quantities in the North Sea occurred at a particularly opportune time both for the UK and for the oil companies. Throughout this century there has been a substantial rate of growth in the world oil industry. Demand for oil, and oil products, has risen steadily as technology in all its forms has been becoming more and more energy (and particularly oil) intensive. Demand for oil has increasingly been met by supplies from the Middle East, especially in the post-war period. For much of this time the oil companies and the oil importing countries have been content with this situation. The Western countries were able to use their superior technology and better bargaining position to ensure stable supplies at a favourable price. This point is illustrated by Figure 1.3, which shows the path of real oil prices from 1900 to 1980.

There has been a continual fall in the real price of oil throughout the whole of the first 70 years of this century. This illustrates the dominant position which the oil companies and the oil consuming countries held. To a large extent once the price of oil had begun to fall it became a self-perpetuating trend. A fall in the real price of oil implies that the best policy which the oil companies can adopt is to develop and deplete their oil stocks as quickly as possible. This is because the alternative of investing in non-oil assets is likely to yield a positive real rate of return as opposed to the negative yield offered by the oil stock. So once the real price of oil begins to fall the normal oil company response is to increase the supply of oil. This increase in supply can well lead to further reductions in the price of oil and so on. Whether or not the continued fall in oil prices can be accounted for on the basis of this market mechanism or as the result of an oligopolistic group of oil companies, the ultimate effect was a substantial shift in wealth away from the oil producing countries and to the oil consuming countries.

Figure 1.3: The Real Cost of Oil 1900-1980 (1970=100)

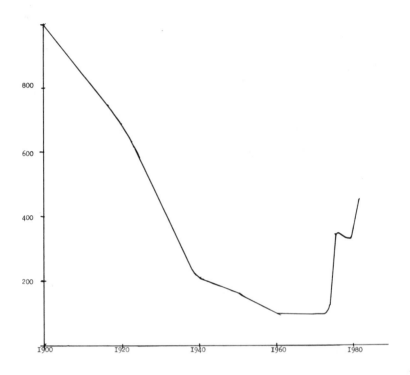

Oil was first discovered in 1859 in Pennsylvania and its commercial use quickly spread through the USA and then to Europe. Initially the industry was characterised by many small producers but a general condition of excess supply and falling prices allowed a single company to attain a dominant position. The company, the Standard Oil Trust, was the main producer until, in 1911, it was forced by anti-trust legislation to split its operations. Of the many companies formed from this move three have particular importance today, Exxon (formerly Esso), Mobil and Socal. These three companies together with Gulf, Texaco, Shell and BP, form the seven major oil companies which have dominated the oil industry throughout most of this century.

During the period before the 1970s the oil industry can be characterised as being in an almost perpetual state of excess supply. The discovery and development of oil in Mexico, Venezuela, Iran and many other areas combined to produce the potential for excess supply. The system of payment to the producing countries, which linked the total payment to production rather than revenue, created the incentive for each individual producing country to maximise its output. This allowed the oil majors steadily to reduce the real price of oil. If an individual country tried to stand against this trend, then the majors could still meet the overall demand from other sources, in particular from the USA oil reserves.

Direct government participation in the oil industry and the oil companies is a long established fact. For example, the British government took a controlling interest in BP (Anglo-Iranian as it was then) in 1913 in order to ensure naval oil supplies. The main move on the part of the oil producing countries to take a direct part in the organisation of the industry did not occur until the late 1930s and 1940s. In 1938, Mexico nationalised all the oil companies in that country following their refusal to improve the conditions of oil field workers. In 1943 the Venezuelan government started a move which resulted in a 50-50 profit-sharing agreement. This system was subsequently adopted in the Middle East by Saudi Arabia, Iraq and Kuwait in 1951. The change in system was an important turning-point as it meant that the host country profited as much from a rise in oil prices as from a rise in production. The producing countries no longer have only conflicting interests, in the sense of competing for market shares in a situation of excess supply, but now they have a common interest in attempting to secure high oil prices.

During the middle and late 1950s there was also the beginning of another challenge to the major oil companies. This took the form of a growth in the number and activity of small independent oil companies as well as an increase in exports from the USSR and Libya. This sudden uncontrolled increase in supply forced the oil majors to reduce oil prices in order to maintain their market share. The response of the producing countries, who now had a common interest in oil prices, was to join together to form OPEC (1960) in order to prevent further price reductions. OPEC was successful in preventing any further price reductions, but was not able to increase the price of oil, due largely to the refusal of Iran to limit its oil production.

The first half of the 1960s was a fairly stable time in the world oil industry. There was a steady decline in North American exports, while its oil imports grew; this was combined with a general increase in dependence on OPEC oil. The events of the late sixties, particularly the Arab-Israeli war in 1967 and the closure of the Suez Canal, brought about an increase in Western dependence on Libyan oil. In 1970 a left-wing government came to power in Libya and immediately reduced oil output and increased the price. Other producing countries quickly followed the Libyan lead and the early seventies saw a succession of small price increases dictated by the oil producers. This culminated in the large increase in 1973/4 when oil prices rose from $2.8 to $11.6 a barrel. OPEC had come effectively to dominate the world oil market.

This long-term trend in the shift in bargaining power and control from the oil companies to the producing countries was important for the development of North Sea oil. It explains why the oil companies were eager to develop the North Sea, despite the fact that even substantial oil finds would have been only marginally profitable at 1970 oil prices, and that the oil companies held no great hope or expectation of making such finds. Sir Eric Drake, Chairman of BP in 1970, has been reported as predicting, in April 1970, that there would not be a major oil field discovered in the North Sea.[1] This was only six months before his own company discovered the Forties field. Indeed, it has been suggested that some of the other major companies (e.g. Mobil and Texaco)

Table 1.2: Exploration and Success Factors for the UK Sector of the North Sea

	Exploration wells drilled	Significant oil and gas finds	Success factors
1964	1	0	0
1965	10	1	0.1
1966	20	4	0.3
1967	42	3	0.07
1968	31	3	0.09
1969	44	6	0.13
1970	22	4	0.18
1971	24	5	0.20
1972	33	6	0.18
1973	42	8	0.19
1974	67	15	0.22
1975	79	27	0.34
1976	58	14	0.24
1977	67	8	0.12
1978	37	3	0.08
1979	33	8	0.24
1980	32	2	0.06
1981	47	12	0.25

Source: Development of the oil and gas resources of the United Kingdom, 1982.

were also becoming disenchanted with the prospects for future major fields. This is illustrated in Table 1.2, which shows the rate of exploration wells drilled between 1964 and 1981.

The exploration rate rose rapidly over the period 1964-7; it then remained fairly stable until 1969 when it dropped dramatically. This was despite the third round of exploration licences issued in 1970. The oil companies were in fact largely continuing their exploration at this time because of their intense need to diversify their interests and production potential outside the OPEC countries rather than because they saw a large potential in the North Sea.

This changed rapidly in the early seventies with two separate developments. Firstly, the discovery of major oil fields established the existence of truly commercial fields in the North Sea. (The Forties field was found in November 1970 (240m tonnes of oil) and then in July 1971 the Brent field was found (229m tonnes of oil).) Secondly, the sudden increase in the world price of oil which occurred in 1973/4 made even the small fields like Montrose (12.1m tonnes of oil) a commercially viable proposition. These two factors combined to produce a major burst in exploration activity during the middle seventies.

Table 1.3 shows the way in which the total stock of discovered and proven oil reserves rose during the 1970s as exploration in the North Sea expanded.

The two large finds in 1970 and 1971 of the Forties and Brent fields led to a surge of exploration throughout the period 1972-6 which met with considerable success. Total discovered reserves rose from a figure of around 500 million tonnes in 1971 (consisting primarily only of the Brent and Forties fields) to approximately 1,800 million tonnes in 1976.

Two reasons are commonly given for the sudden fall off in the rate of discoveries from 1977 onwards. The first of these is the change in the tax structure which occurred over this period and which made oil exploration much less profitable. This factor will be discussed later, however, as it is probably not the major cause. The second factor is the uneven pattern of licensing which successive governments have generated.

There are two basic types of licences: exploration licences and production licences. Despite their titles any actual drilling requires a production licence and the exploration licence only allows an area to be given a general geological survey. To date the production licences have been issued in seven licensing rounds between 1964 and 1980. Table 1.4 gives details of the dates of each round, the number of

Table 1.3: Fields Discovered and Proven. Recoverable reserves (m. tonnes)

Year	Field discovered	Total oil dis-covered	Cumulative total
1968	—	—	—
1969	Montrose	12.1	12.1
1970	Forties	240	252.1
1971	Brent, Auk, Argyll	229	481.1
1972	Beryl, S. Cormorant	94	575.1
1973	Piper, Dunlin, Thistle Alwyn, Hutton, Maureen	284	859.1
1974	Ninian, Claymore, Magnus, N. Cormorant, Andrew, Buchan, Statfjord	443	1,302.1
1975	Murchison, Brae, Tartan, Fulmar, N.W. Hutton, Tern, Lyell, Crawford, Heather, Beryl N., Thistle N.E.	411	1,713.1
1976	Beatrice, Thelma, Renee	100	1,813.1
1977	Clare, Brae	—	
1978	Clyde	—	
1979	Tiffany		
1980	N.E. Brae	—	—

Source: Development of the oil and gas resources of the United Kingdom, 1982.

licences which were awarded and the number of 'blocks' which these licences covered. A 'block' is the actual area of the continental shelf which is open for drilling, each block representing approximately 100 square miles.

The pattern of the allocation of licences on the part of the government has been very erratic. In the first two rounds virtually the whole available North Sea area was offered for licensing. A large gap then occurred between the second and third round while most of the exploration for natural gas was undertaken. In the third round a much smaller number of blocks was offered although the number of final awards was not very much less than the second round. The fourth round occurred fairly quickly after the third, and many more blocks were both offered and issued (largely because of the Forties discovery). There was then no further award until the fifth round five years later. A comparison of Table 1.4 and Table 1.2 shows that the exploration drilling which occurred as a result of the fourth round of licences did not peak until 1975. This illustrates the long delays which can sometimes be involved

Table 1.4: The Licensing Pattern for the UK Continental Shelf

Round date	No. of blocks offered	No. of blocks awarded	No. of licences
1 1964	960	348	53
2 1965	1102	127	44
3 1970	157	106	61
4 1971/2	436	282	118
5 1976/7	71	44	28
6 1978/9	46	42	26
7 1980/81	80 + 20[a]	90	90

Note: a. The extra blocks represent an intent to award 20 company nominated blocks. That is 20 blocks specified by companies rather than the government.
Source: Development of the oil and gas resources of the United Kingdom, 1982.

in exploratory drilling. Not only must the site be chosen and financial backing be provided but the number of exploratory wells drilled is limited by the number of mobile rigs available. The construction of new rigs often takes two to three years and the only alternative to this is to move rigs into the area from some other offshore site. This can only be done gradually as rigs become available. The movement of mobile rigs into the UK area of the continental shelf is detailed in Table 1.5.

The build-up of the number of rigs, after the fourth licensing round, is evident. The fact that rig activity began to fall off in 1976 is not at all surprising; it is due mainly to the fact that the bulk of the fourth round exploration had been completed and that there had not been a fifth round issued early enough to allow a constant level of activity. The announcement of the fifth round in 1976 caused a slight reverse in the decline in 1977 but as the licensing round was so small the decline in activity continued in 1978 and 1979. The government realised around the time of the fifth round that the large fourth round, with a long subsequent gap, had been a mistake. It was felt that it generated too much exploration in a large block which would ultimately lead to very high levels of oil production in the early 1980s followed by a very fast decline in production as there are relatively few further discoveries to come on stream.[2] It was recognised, therefore, that it would be better to maintain a lower but steadier exploration rate. This could be achieved by a policy of much more frequent smaller licence rounds. It is this policy which has generated rounds five, six and seven. It is to be hoped that these exploration rounds will lead to a lower but steadier rate of oil production in the late 1980s.

Table 1.5: Mobile Rig Activity on the UK Continental Shelf (in rig years)

	1970	1971	1972	1973	1974	1975	1976	1977	1978	1979	1980	1981
East of England	3.1	1.2	1.9	2.5	1.7	1.6	1.5	1.4	0.6	0.5	—	1.2
East of Scotland	2.2	3.2	4.1	3.8	8.2	12.0	8.4	9.5	9.3	9.6	12.8	16.3
East of Shetland	—	0.8	2.7	6.9	12.4	13.6	9.9	10.1	5.9	4.5	6.3	6.5
West of England/ Wales	—	—	—	0.1	0.7	0.2	1.0	0.8	1.1	0.2	—	0.2
West of Shetland	—	—	0.1	—	1.5	0.3	0.4	1.8	1.1	0.4	1.5	0.4
Channel and SW Approaches	—	—	—	—	—	—	—	—	0.1	0.9	—	—
Total all areas	5.3	5.2	8.8	13.3	24.5	27.7	21.2	23.6	18.1	16.1	20.6	

Source: Development of the oil and gas resources of the United Kingdom, 1982.

Figure 1.4: The UK Continental Shelf

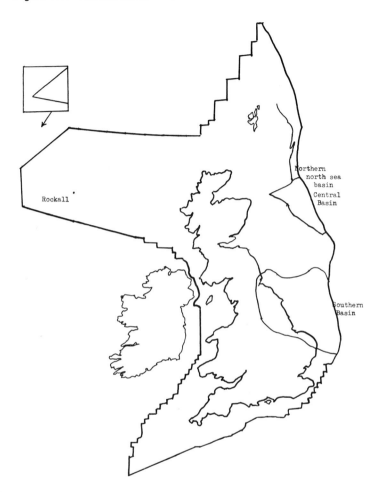

The term North Sea oil is often used rather loosely to mean the offshore oil deposits of the United Kingdom. This can lead to a considerable degree of confusion especially when we later examine a number of varying estimates of the size of the oil resource. Figure 1.4 shows the present boundaries of the designated area of the UK continental shelf. The vast majority of the early gas finds occurred in the southern part of the North Sea in a geological formation called the Southern North Sea Basin. The early oil discoveries were made in an area called the

Central North Sea Basin which extends up to the two small fields Tiffany and Toni. The area to the west of this is the Moray Firth Basin and to the north the East Shetland Basin. These three basins contain almost all the discovered offshore oil reserves. Both the Central North Sea Basin and the East Shetland Basin extend well beyond the UK boundary into the Norwegian, Danish and German areas. These three basins are sometimes collectively referred to as the Northern North Sea Basin, and when oil reserves are attributed to this area it will generally exclude the non-UK areas. There are principally two areas of confusion, therefore; North Sea oil may well imply areas and oil reserves outside UK jurisdiction, and UK North Sea oil will not normally include very large areas of the UK continental shelf such as the Rockall Trough or the English Channel.

The impact of oil has been felt in three main areas. Firstly, there is the direct increase in national income due to oil output adding to the Gross Domestic Product (GDP). Secondly, there is the major impact which domestic oil development and production will have on the balance of payments. Thirdly, there is the general macroeconomic effect of large government tax revenues flowing from the oil sector as well as the direct economic effects of investment and employment.

The direct GDP contribution of the North Sea is relatively straightforward. The last two rows of table 1.6 show the rapid rise in the contribution of the North Sea to the GDP over the period 1974 to 1980. While the direct contribution in 1980 is very large, nearly £9 billion, it is obvious that the North Sea sector is in no sense a dominant sector of the British economy. This £9 billion represents less than 5 per cent of the total GDP figure for the UK. So that while the North Sea contributes almost twice as much as agriculture, forestries and fisheries this is only about one sixth of the contribution of total manufacturing and about half the contribution of the insurance and banking sector. In the UK the oil sector can never have an overwhelmingly dominant effect on the economy simply because it will always be a relatively minor part of general economic activity.

Second, the balance of payments benefits greatly from the development of oil. After 1972 the current account on the UK balance of payments became increasingly negative. The initial move into deficit was due to the general economic climate and cannot be connected with the oil sector directly. However, the sudden increases in the price of oil which occurred towards the end of 1973 and throughout 1974 greatly exacerbated the balance of payments deficit. Table 1.7 illustrates that while the physical quantity of imported oil fell considerably

Table 1.6: Direct Contribution of Oil and Gas to the Balance of Payments (£ bn current prices)

	1973	1974	1975	1976	1977	1978	1979	1980	1981
Value of sales									
Gas	0.13	0.17	0.19	0.26	0.32	0.43	0.54	0.65	0.84
Oil	0.002	0.003	0.06	0.65	2.23	2.81	5.69	8.85	12.34
Total	0.132	0.173	0.25	0.91	2.55	3.24	6.23	9.50	13.18
Less									
Net imports of goods and services	−0.16	−0.33	−0.82	−1.18	−1.23	−0.73	−0.61	−0.61	−0.78
Less									
Interest, profits and dividends due abroad	−.009	−.01	−.023	−.01	−0.55	−0.74	−1.37	−2.22	−2.35
Current account Impact	−.037	−.167	−.593	−.294	0.77	1.77	4.25	6.67	10.05
Capital account impact	0.07	0.23	0.95	1.14	1.51	0.79	0.70	0.84	1.66
Direct balance of payments effect	0.033	0.063	0.357	0.864	2.28	2.56	4.95	7.51	11.71
Adjusted GDP contribution (From Table 1.1)	.02	.22	.32	1.03	2.87	3.72	6.52	9.8	
As % GDP	.3	.3	.3	.8	2.0	2.3	3.4	4.4	

Source: UK Balance of Payments, 1982 Edition, p. 47.

Table 1.7: The Quantity and Value of Imported Oil, 1970-1980

	1970	1971	1972	1973	1974	1975	1976	1977	1978	1979	1980
Quantity of oil imports (mn tonnes)	100.7	107.3	104.3	113.2	110.8	87.1	86.9	68.5	65.4	57.8	44.7
Value of oil imports (£ million)	686.9	929.6	914.0	1296.2	3725.6	3371.3	4444.7	3970.7	3505.4	3671.3	4292.0
Average cost of imported oil (£/tonne)	6.81	8.66	8.76	11.44	33.62	38.67	51.11	57.91	53.56	63.42	95.84

Source: Digest of UK energy statistics.

between 1970 and 1980, the cost of these imports rose enormously. It should perhaps be pointed out that there is no contradiction between being self-sufficient in oil in 1980 and importing 45 million tonnes in 1980. Because oils from different regions vary enormously in terms of their chemical properties and qualities, it is desirable that a large part of our oil is exported so that oils of different quality may be imported in its place. So net self-sufficiency implies that we import as much oil as we export; it does not imply zero gross oil imports.

The trend which had persisted from 1971 to 1974 of a continually worsening balance of payments situation changed in 1975 and the situation improved dramatically until 1978 when a sizeable surplus existed. There was a slight reverse in 1979, due to a large increase in imports, followed by a very large surplus in the recession year 1980. It is, of course, impossible accurately to describe what the situation would be in the absence of oil. The exchange rate would obviously be much lower, non-oil imports and domestic consumption would have had to be lower and the general economic climate would have been radically altered. It is, however, possible to assess the direct contribution which oil has made to the balance of payments. Figures for this effect are presented in Table 1.6 for the period 1973-81. The direct import saving effect of the North Sea is shown in the first row. These figures rise very rapidly as oil production rises from 1975 through to 1981 (see Table 1.8 for total oil production and a field by field breakdown). These figures do not, however, represent a total gain to the balance of payments as a certain amount of investment goods had to be imported and there was also a flow of funds abroad due to interest, profits and dividends (IPD). When these effects are removed the impact on the current balance is shown as row four. This still will not represent the full balance of payments effects as there have been large flows of capital into the country in order to finance North Sea development. These flows are shown in row 7 as the capital account impact. The sum of the capital and current account effects gives the overall impact on the balance of payments. The impact of the North Sea on the balance of payments, particularly in the last three years of the period, can be seen to be immense. The current balance in 1981 actually stood at a figure close to £6 billion. Table 1.6 shows that if the North Sea sector had simply disappeared, with no other economic changes taking place, then this surplus would have been transformed into a deficit of £6 billion. Perhaps the greatest practical benefit which the North Sea sector has produced is to protect the UK economy from having to make the very great structural adjustments which such large and persistent deficits would have induced.

Table 1.8: Oil Production from the UK Sector of the North Sea and UK Oil Consumption

Field	Start-up time	Production						
		75	76	77	78	79	80	81
	(Years)	(million tonnes)						
Argyll	4	0.5	1.1	0.8	0.7	0.8	0.8	0.5
Auk	5		1.2	2.3	1.3	0.8	0.6	0.6
Beatrice	5	—	—	—	—	—	—	0.2
Beryl	4		0.4	3.0	2.6	4.7	5.4	4.7
Brewt	5		0.1	1.3	3.8	8.8	6.8	11.1
Buchanan	7	—	—	—	—	—	—	0.9
Claymore	3			0.3	3.0	4.0	4.4	4.5
S. Cormorant	7					0.04	1.1	0.7
Dunlin	5				0.7	5.7	5.2	4.7
Forties	5	0.6	8.6	20.1	24.5	24.5	24.6	22.8
Heather	5				0.1	0.8	0.7	1.2
Montrose	6		0.1	0.8	1.2	1.3	1.2	1.1
Murchison		—	—	—	—	—	0.4	3.1
Ninian	4				0.04	7.7	11.4	14.3
Piper	3		0.1	8.0	12.2	13.2	10.4	9.8
Statfjord	5					0.04	0.5	1.2
Tartan	6	—	—	—	—	—	—	0.7
Thistle	5				2.6	3.9	5.3	5.5
Average start-up time	5							
Total production[1]		1.6	12.1	38.3	54.0	77.9	80.5	89.4
Consumption		93.3	92.5	92.9	94.0	94.0	80.8	74.4

Note: 1. Including onshore production 92.9.
Source: Development of the oil and gas resources of the United Kingdom, Digest of United Kingdom Energy Statistics.

The final, main channel through which North Sea oil will affect the macroeconomy is through the government revenues which it will generate. This effect was not a major factor in the economy during the 1970s. This is because the taxation structure allowed the oil companies to reclaim much of their capital expenditure before petroleum revenue

Table 1.9: Government Revenues 1976-80 (£million)

Year	Royalties	PRT	Corporation tax	SPD	Total
1976/7	71	0	10	—	81
1977/8	228	0	10	—	238
1978/9	289	183	90	—	526
1979/80	628	1435	266	—	2324
1980/1	940	2420	480	—	3840
1981/2	1350	2380	650	2050	6430

Source: Development of the oil and gas resources of the United Kingdom, 1982.

Table 1.10: Capital Expenditure in the North Sea, £million (constant 1975 prices)

1972	211	1975	1371	1978	1577
1973	327	1976	1843	1979	1294
1974	691	1977	1699	1980	1229

Source: National Income and Expenditure, 1981.

tax had to be paid. The details of the tax structure will be left to the next chapter but Table 1.9 shows how actual government revenue has risen over the period 1976-81. In the last three years total revenues have been mounting quite substantially, but they do not approach the large amounts which should come to hand during the early and middle 1980s.

Investment directly in the North Sea has been considerable over the period 1972-80.

Table 1.10 shows the pattern of direct capital expenditure in the North Sea. Capital expenditure peaked in 1976 and then remained fairly stable for three years, and began to decline in 1979/80. In the early years only a small part of total capital expenditure was spent within the UK (an approximate estimate might be 25-35 per cent). Under direct government pressure this has changed and the majority of the work is now being undertaken in the UK, 66 per cent in 1978, 79 per cent in 1979 and 71 per cent in 1980. Investment figures of this magnitude are bound to have an effect on the overall level of demand as well as a much stronger sectoral effect. The direct offshore employment effects are also significant, despite the fact that the industry is primarily a capital-intensive one. Approximately 21,000 people were employed

Table 1.11: Operating Costs for the North Sea Oil Fields

	Total operating cost (£m 1980)	Cost per tonne of product (£m 1980)
1976	135.8	11.7
1977	236.2	6.3
1978	345.0	6.5
1979	483.6	6.3
1980	582.0	7.4
1981	773.0	8.7

Source: Development of the oil and gas resources of the United Kingdom, 1982.

directly on offshore work in 1981 and of these 85 per cent were UK residents.

Table 1.11 shows the pattern of direct operating costs in the North Sea oil fields from 1976-80. This table shows an almost perfect U-shaped average cost curve; the early production in 1976 showed a relatively high average cost. This cost fell quite rapidly as output rose in 1977, then remained steady throughout 1978-9. In 1980 the average cost began to rise quite sharply as the smaller, less profitable, fields such as Murchison came on stream and production began to fall in some of the older fields such as Piper, Brent and Auk. The average cost of production overall is likely to continue to rise as smaller fields are exploited and the output of larger fields such as Forties begins to decline.

Notes

1. In C. Callow, *Power From The Sea*, Gollancz, 1973.
2. See *Financial Times*, 2 April 1981, 'Doubts on North Sea Depletion Plans' by Ray Dafter. This article demonstrates the government's need to smooth out the coming peak in oil production.

2 PROSPECTS FOR OIL AND GAS

Introduction

This chapter examines the likely future developments of oil and gas from 1981 onwards. It will be necessary to examine and compare a number of different estimates and consider where they diverge in terms of assumption and methodology. First is the question of what the total stock of oil reserves might be; secondly the likely production profile; and finally the pattern of total revenues accruing to the government. None of these three things can be accurately assessed. The total stock of oil is, at the most basic level, simply unknown. The production and government revenue patterns both depend on secondary factors such as future exploration and development rates or the trends in world oil prices.

Total Reserves of Oil and Gas on the UK Continental Shelf

The Oil Sector

Before estimates of oil and gas reserves are present we must explain what is meant by the term reserves. The amount of oil which will be recovered from a field is dependent on three quite separate factors. The first is geological: how much oil is in the field, the type of oil and rock of which the field is made up, and any geological faults or anomalies which exist. The second main determinant is the price of oil and the tax structure to which the field is subject. Production from a field will ultimately be stopped when the marginal cost of production rises above the marginal revenue. The final set of factors is technological, and includes the actual method used to exploit the field and any secondary techniques used to enhance the overall recovery rate. A change in the technique used to exploit a field may drastically increase the amount of oil recovered, particularly in offshore oilfields. In the case of a field which is not exploitable using conventional techniques, the introduction of a new method of exploitation may raise that field's recoverable reserves from zero to a large amount. Enhanced recovery also has a major effect on the quantity of oil recovered. The average oilfield will only produce about one-third of the total oil in the ground. The remaining oil is left locked in the rock. This may be increased by such techniques as re-injecting natural gas or injecting steam. Ultimately it is hoped that recovery rates might be in the range of 50-70 per cent, which would double the figure for recoverable oil reserves.

Table 2.1: Department of Energy's Estimates of the Oil and Gas Reserves on the UK Continental Shelf (as at 31 December 1981)

| | Oil (million tonnes) | | | |
	Proven	Probable	Possible	Total
Total reserves	1050	575	675	2,300
Estimated range of discovered reserves				(1,475-1,775)
Cumulative production				354
Potential undiscovered reserves				275-2,175
Estimated reserves originally in place				2,100-4,300
	Gas (million tonnes oil equivalent)			
Total discovered reserves	553	285	331	1169
Estimated range of discovered reserves				(733-979)
Cumulative production				348
Potential undiscovered reserves				54-550
Estimated reserves originally in place				1,125-1,875

Source: Development of the oil and gas reserves of the United Kingdom, 1982.

When figures for oil reserves are presented they are based on a wide variety of assumptions about these three factors, and are subject to a considerable degree of uncertainty. To allow for this uncertainty reserve estimates are often divided into three sectors: Proven, Probable and Possible reserves. The general definition of these three terms is that Proven reserves are virtually certain to be technically and economically producible, Probable reserves are assessed to have a better than 50 per cent probability of being technically and economically produced, Possible reserves have a significant but less than 50 per cent probability of being technically and economically produced.

The starting place for an assessment of total oil and gas stocks must be the official estimates. Table 2.1 presents the Department of Energy's figures for total UK oil stocks. The immediately striking thing about this table is the very wide range given for the total reserves figure. The stock of total oil reserves originally in place is estimated at 2,100-4,300 million tonnes of oil. The range presented for discovered reserves is smaller than the range of the proven and total possible discovered reserves as the Department of Energy argue that the extreme ends of this range are unlikely to represent reality. This reduced range is then added to total cumulative production and the estimated range of potentially undiscovered reserves to give the overall total estimate. The development of the Department of Energy's total oil stock estimates since they were first published in 1975 reveals an interesting pattern. In 1975 the total oil resources of the UK continental shelf were estimated at 4,500 million tonnes. During the next three years the published estimate remained constant at a range of 3,000-4,500 million tonnes. In 1979 the lower end of the range was reduced substantially while the upper end had a slight reduction to 4,400 million tonnes. From this point onwards both ends of the range have been gradually reduced each year to the current figure. This pattern is rather surprising in view of the very large increases in the price of oil which occurred in 1979/80 and the trend towards higher recovery rates, both of which should have led to a considerable increase in the quantity of recoverable reserves. It would seem, therefore, that either the early estimates were unrealistically high or that more recent estimates have failed to rise in the way we might expect.

The figure of 2,100-4,300 million tonnes of oil is, despite these reservations, still a very considerable one and it is worth pointing out that at 1980 consumption levels (80.8 million tonnes) self-sufficiency could be maintained for 16-53 years. Moreover, these figures are very conservative when compared to most other estimates. Professor C. Robinson[1] gives a general consensus figure for total UK North Sea reserves, between 55°50′ North and 62° North, in effect the main area of existing oil discoveries, of 2,900-3,300 million tonnes. This figure cannot be compared exactly with Table 2.1, because the areas being considered do not exactly coincide, but the Department of Energy figure for reserves in the unlicensed parts of the continental shelf (100-1,050) can be added to Robinson's figure without fear of double counting (although some areas may not then be counted at all). This process gives a total reserve figure of 3,000-4,350 million tonnes. The upper figure is close to the Department of Energy range but the lower figure is considerably higher and would represent 37 years of production at 1980 self-sufficiency level.

Professor Robinson's figures represent the general views of the oil industry fairly well but there is another faction in the debate, represented perhaps best by Professor P. Odell. Professor Odell has estimated[2] the total stock of recoverable oil in the area which he terms the North Sea basin (that is the central North Sea basin, the East Shetlands basin and the Moray Firth basin) to be between 6,400 and 18,600 million tonnes. This figure is not comparable with Professor Robinson's figure as it includes the Norwegian sector of the North Sea basin. Well over half the total area of the basin, as defined by Odell, lies in the UK sector so it is a fair process to allocate half Odell's estimated reserves to the UK. This would give a figure, comparable with Professor Robinson's, of 3,200-9,300 million tonnes. When the figure (from the Department of Energy) for non-North Sea basin reserves is added, the total oil stock would be 3,300-10,350 million tonnes. This would represent self-sufficiency, at 1980 consumption levels, for a period of 41-128 years.

The fact that the estimates of Robinson and Odell are so different might lead one to suspect that they had used radically different approaches or methodologies. This is not in fact the case. Both have followed the same approach. Oilfields are divided into a number of classes by order of size, a probability of finding a field of each size is assumed (largely arbitrarily); this probability will change as exploration takes place. The number of exploratory wells drilled each year is then decided, this figure multiplied by the probability of finding a given size field gives the amount of oil found in each year and so the total oil stock is derived. Robinson and Odell differ in two basic assumptions; Robinson assumes that only 212 exploratory wells will be drilled between 1977 and 2000 and that the probability of field discoveries will fall off quite quickly over this time period. Odell assumes a total of 1,375 exploratory wells drilled over the period and that the probability of successful oil discoveries remains stable over a much larger search pattern. The assumption about the rate of exploration is, in fact, not important, it cannot change the total stock of oil but merely affects its rate of discovery. The crucial assumption is the one about how the probability of discovery falls off as drilling proceeds. There is no way to assess these assumptions except in a historical context after the oil has been discovered. The assumptions in effect rest on the authority of the respective authors. It can, of course, be argued that there is no real alternative; in the face of uncertainty all that can really be done, at a fundamental level, is to give an informed guess. There is one alternative approach, however. Barouch and Kaufman (1976) have outlined a fairly simple model of the discovery rate of natural resources such as oil and gas. Their basic idea is that reserves are likely to be lognormally distributed in any given geological setting and that the probability of finding any given reserve, or oilfield, is proportional to the size of the field. The discovery rate will therefore be the convolution of a lognormal distribution and the proportional probability of discovery. This in effect says simply that the largest reserves will generally be found first and that further search will yield increasingly less and less returns. In effect we have a fairly simple concept of a diminishing marginal product curve with respect to search activity in a given area. W.D. Nordhaus has pointed out that a simple functional form which closely fits the Barouch and Kaufman analysis is a negative exponential one of the following form

$$q' = \alpha e^{-BS}$$

where Q is total oil discovered, S is the total of search which has been undertaken, α and B are parameters and $q' = \frac{\delta Q}{\delta S}$ the marginal product of search (i.e. oil discovered).

The total quantity of the reserve which is available for exploitation will therefore be equal to the total area under this curve. That is

$$Q = \delta^{\infty} \alpha e^{-BS} dS = \frac{\alpha}{\beta}$$

Clearly before this formula can be put into a practical form a number of changes need to be made to it. The concept of search used so far has been kept deliberately vague. Nordhaus used the concept of actual square footage drilled. This is in many ways an ideal figure but it is difficult to apply to an offshore site such as the North Sea and such data is simply not available. A good proxy for the amount of search which has occurred at any point in time in the North Sea would be the number of exploratory wells drilled. The marginal return to search is therefore defined as the number of barrels of recoverable oil discovered per well drilled in a time period. The only remaining point is to express the functional form as a discrete one and to transform it into a form suitable for estimation. A discrete version of the model is

$$q' = \alpha(1+\beta)^{-S}$$

where S is now the number of explorative wells drilled. Taking logarithms of this gives

$$\ln q' = \ln \alpha - (\ln (1+\beta)).S$$

So if the equation

$$\ln q' = \eta + \phi.S$$

is estimated, $\eta = \ln \alpha$ and $\phi = \ln(1+\beta)$. The data set used is given in Table 2.2.

Table 2.2: Data for the Estimation of Total Oil Stocks

Year	UK exploration wells drilled	Cumulative total	Total oil discoveries (BM)	Oil disc/well drilled (BM)
1970	22	22	2,280	103
1971	24	46	2,296	95.6
1972	33	79	760	23
1973	42	121	2,408	57
1974	67	188	2,788	41
1975	79	267	1,834	23
1976	58	325	810	13

Source: Wood Mackenzie.

This data gives only 7 observations; this is obviously a low figure but there seems to be no practical way to increase it. It is not possible to extend the set either before 1970 (because there was so little exploration done then) or beyond 1976 (because comprehensive figures for discovery sizes are not yet available). Similarly an attempt was made to construct quarterly figures but this failed because the recorded discovery dates tended to fall in the last quarter of each year and so there were many quarters where the discovery rate was zero.

The equation was estimated using a maximum likelihood technique[3] which has been especially developed for small sample sizes. The estimated equation was

$$\text{Ln } q' = 4.433 - 0.00512.S$$
$$\quad\quad\quad (28.470) \;(-5.21) \quad\quad \text{(t Ratios)}$$
$$\bar{R}^2 = 0.755$$

which gives an original equation of

$$q' = 84.2 \,(1 + 0.005133)^{-S}$$

Both parameter estimates are of the correct sign and are significant at a 99 per cent probability level. The estimated equation is compared with the observed points in Figure 2.1.

Given these estimates it is now possible to estimate the stock of oil reserves implied by the equation:

$$Q = \frac{\alpha}{\beta} = \frac{84.2}{0.005133}$$

= 16,402m barrels or 2187 million tonnes.

This figure does not, however, represent the stock of oil in the whole of the UK North Sea basin, because search has been restricted to those areas of the basin which are under licence. The North Sea basin comprises approximately 870 blocks, about 130 of which remain unissued. On the assumption that the blocks are homogeneous to a reasonable degree it is possible simply to multiply the oil stock figure by a factor of 870/740 to derive an estimate of the basin's total reserves. The figure reached is 2,571 million tonnes. This figure is reasonably in line with the Department of Energy (2,000-3,250) and Professor Robinson (2,900-3,300), but it is below Professor Odell's figure (3,200-9,300). It would seem, therefore, that Professor Robinson's assessment of discovery probabilities is more realistic than Professor Odell's.

In assessing total oil stocks it would obviously be a mistake to concentrate solely on the North Sea basin. The Rockall trough is an area equal in size to the North Sea basin and it has some very good geological prospects. Similarly, the English Channel and Irish Sea combine to make an area nearly as large as the North Sea basin. If we made the simple but tenuous assumption that these areas turn out to be as fruitful as the North Sea it would effectively triple our reserves to a figure in the region of 6,000 million tonnes. This would allow self-sufficiency of oil at 1980 consumption levels for 85 years. It is not of course possible to justify this assumption; until considerable exploration is undertaken no estimates can be presented with any confidence. However, this consideration makes it clear that the Department of Energy's figure of 100-1,050 million tonnes in the unlicensed areas (including the unlicensed North Sea basin) may be very conservative.

In conclusion, a figure high in the Department of Energy's range

Figure 2.1: Comparison of Predicted and Actual Oil Discovery Rates
for the North Sea

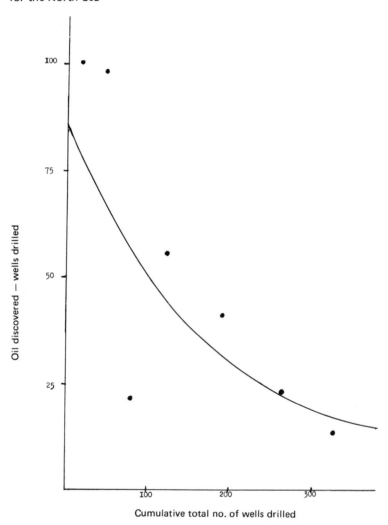

Cumulative total no. of wells drilled

for total oil reserves (2,100-4,300 million tonnes) seems reasonable as
a conservative estimate. A point estimate of 4,000 million tonnes would
yield self-sufficiency (at 1980 levels) for nearly 50 years. The stock of
oil in the whole UK continental shelf area is therefore a very large
energy resource.

It may be argued that we should base our economic planning only on absolutely certain oil reserves, in effect the lower figure given by the Department of Energy. This approach can, however, lead to completely inappropriate long-term strategies as they will be founded on the assumption that no further oil whatsoever will be discovered. This assumption is clearly wrong. Policy should be founded on the assumption that the continental shelf hydrocarbon deposits are large and will last well into the twenty-first century.

The Gas Sector

The reserve figures for natural gas are somewhat less controversial, simply because less non-government research has been done on them. The Department of Energy's figures are presented in the lower half of Table 2.1.

In 1980 total gas consumption was 41 million tonnes oil equivalent. At this consumption level the gas stock will last from 27 to 45 years. Most of the arguments for ignoring the lower end of the Department of Energy's range presented in the last section apply here equally. So it is reasonable to expect total gas stocks to be able to maintain 1980 consumption levels for 35 to 45 years.

The Rate of Future Oil Extraction

The previous section has dealt with the total stock of oil and gas in the UK continental shelf. This information is important and has profound long-run implications, but it says nothing about the levels of oil production over the next decade. Indeed, as long as oil production does not exhaust the total stock, the level of oil stocks is irrelevant to production questions. When actual future production rates are being considered there are two distinct time periods which must be thought of separately. Over a time horizon of less than five years there is no likelihood of actual production rates being influenced by fields which are not yet discovered. To estimate such a short-term production profile all that needs to be done is to calculate the sum of expected production in all the existing fields and those under development. To predict the pattern of production beyond this five-year period is a much more uncertain exercise, requiring assumptions both about the rate at which future discoveries are made and the rate at which they will be exploited. The degree of uncertainty which exists even within this five-year period is still, however, quite considerable. Production

Table 2.3: Government Forecasts of Production Compared with the
Actual Out-turn (million tonnes)

Forecasts made in:	1976	1977	1978	1979	1980	1981
1975	17½	40	62½	85-95	100-130	125-160
1976	15-20	34-45	55-70	75-95	95-115	—
1977		40-45	60-70	80-95	90-110	100-200
1978			55-65	80-95	90-110	100-120
1979				70-80	85-105	95-115
1980					80-85	85-105
1981						80-95
Out-turn	12.2	38.3	54.0	77.9	80.5	89-4

Source: Economic Progress Report, March 1982.

rates, even only a short period ahead, can be greatly affected by technical problems, delays in field start-ups or disappointing field performance and a wide range of other factors. Table 2.3 illustrates the Department of Energy's oil production forecasting record.

The general pattern of estimates and out-turns is that the initial estimate is very high, the estimate then falls over time, but even in the year immediately before production the forecast is often well above the actual out-turn.

There are a number of sources of future production estimates: the Department of Energy, the National Institute, other economic forecasting institutions, various stockbrokers, in particular Wood Mackenzie, as well as individual forecasters. Some of these estimates are compared in Table 2.4.

The immense uncertainty of the projections which go beyond 1985 is perhaps best illustrated by a breakdown of the Rowland figures. Table 2.5 shows the division of total production into three classes: established commercial fields, potentially commercial fields, and new discoveries. In the early years the estimate for total production is derived from the relatively certain established fields but as the forecasts move further into the future, first the potentially commercial fields then the new discoveries become increasingly important.

Table 2.5 shows only one possible view of the development of the North Sea. The production profile for the established commercial fields is based on field-by-field current production plans. The second two categories are based on either current production plans or assumptions about likely development. This field-by-field approach has, in

Table 2.4: Estimates for Future Oil Production (million tonnes)

	C. Rowland	Dept. of Energy	National Institute	Phillips & Drew	Wood Mackenzie
1982	112	90-105	100	109	108
1983	122	90-115	105	119	116
1984	128	95-125	105	136	126
1985	129	95-130	110	133	123
1986	127		110		114
1987	126				102
1988	129				89
1989	137				75
1990	138				64

Sources: Department of Energy, March 1982; National Institute: National Institute Economic Review, November 1981; Wood Mackenzie: North Sea Service; C. Rowland: University of Surrey; Phillips & Drew: November 1981 Economic Forecast.

Table 2.5: The Breakdown of the Rowland Production Estimates (million tonnes)

	Established commercial fields	Potentially commercial fields	New discoveries	Total
1982	112	0	0	112
1983	122	0	0	122
1984	128	0	0	128
1985	126	3	0	129
1986	114	13	0	127
1987	102	23	1	126
1988	88	39	2	129
1989	75	56	6	137
1990	63	63	12	138
1995	27	28	68	123
2000	7	10	77	94

the past, tended to overestimate actual production figures (see Table 2.4) as insufficient weight has been given to the magnitude of delays and general production shortfalls. The possible effects of a depletion policy have also not been taken into account. Figure 2.2 shows the Department of Energy's estimate of oil production to the year 1990. These projections are universally lower than the Rowland figures, and this illustrates the large degree of uncertainty which exists in forecasting this type of variable.

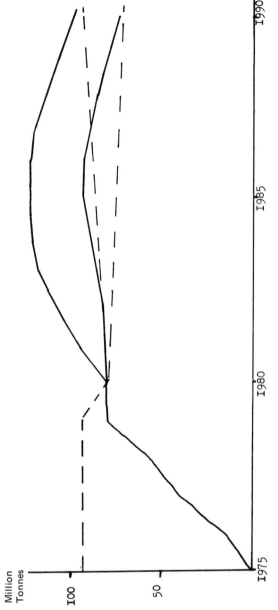

Figure 2.2: The Dept. of Energy's View of Possible Oil Production and Demand Levels

Source: Minutes of Evidence before the Select Committee on Energy (HC 321 1981)

The Rate of Future Gas Extraction

In comparison with oil there are relatively few forecasts of future gas production. There seem to be two reasons for this; first, the gas sector does not generate large tax revenues so it is less important to the large economic models, e.g. CEPG and the LBS do not give any gas forecasts. Second, the gas sector has passed its phase of rapid expansion and has now reached a plateau of stable production which means that over the medium term future gas output is of relatively less significance. In 1980 total gas production was 32 million tonnes oil equivalent; this rate of production is likely to remain fairly stable throughout the 1980s although there may be a slight trend towards increased production. In the longer term production levels in the late 1980s and the 1990s will depend on decisions which are yet to be made and there is considerable uncertainty as to whether output in the 1990s will rise or fall.

Revenues from the Oil Sector

The government revenues which are generated in the oil sector depend obviously on the structure of the prevailing tax system. It is useful, therefore, briefly to review the development and objectives of this system in the UK before considering detailed revenue estimates. Under the terms of the 1964 Continental Shelf Act revenues were to accrue to the government via two distinct channels, a royalty system comprising 12½ per cent of the value of all oil produced and normal corporation tax which would have taken approximately 50 per cent of the oil companies' profits. This taxation system was enormously favourable to the oil companies and partly explains their willingness to embark on costly exploration in the period 1968-72 when the prospects for large finds were highly uncertain. From 1972 onwards the scale of North Sea oil discoveries became increasingly obvious and there was mounting pressure to alter the taxation system to one which recognised the huge potential which had been discovered. These pressures brought about the Oil Taxation Act (1975) which altered the corporation tax system by setting up a 'ring fence' around the North Sea which stopped the companies from setting North Sea profits against losses made elsewhere in the world. This act also introduced a new tax called Petroleum Revenue Tax, which was a tax on total revenue, at a rate of 45 per cent, after various allowances were made for capital expenditure and an initial annual oil allowance. In 1979 it was realised that despite these changes the oil companies were still in a highly favourable position, largely because of the recent increases in oil prices, and as a result the capital and oil allowances were reduced and the rate of PRT was increased to 60 per cent. In March 1980 the rate of PRT was again raised, this time from 60 per cent to 70 per cent and changes were made in the payment system so as to bring forward the date of PRT payments. In the March 1981 budget further restrictions on the PRT allowances were announced as well as a new tax, a Supplementary Petroleum Duty (SPD), to be charged at the rate of 20 per cent on gross oil revenues. So revenues then flowed through four distinct channels, Royalties (12½ per cent), SPD (20 per cent), PRT (70 per cent) and Corporation Tax (52 per cent). This made for a highly complex system which was difficult both to operate and to forecast. The oil companies particularly objected to the supplementary petroleum duty as this tax was levied on pure revenues and therefore tended to make many of the smaller fields unprofitable. PRT acts like a profit tax rather than a revenue tax due to the capital allowances. In the March 1982 budget it was announced that SPD would be abolished at the end of 1982, but be replaced by the advanced payments of PRT (APRT) at a rate of 20 per cent of gross revenues after an allowance of 1 million tonnes of oil per field. This tax will differ from SPD in that it will be set off in full against ordinary PRT liabilities. It was also announced in the budget that the rate of PRT taxation would be increased to 75 per cent from 1 January 1983.

Forecasting future tax revenues involves not only modelling this complex structure of taxes but also forecasting five separate trends. These are: future world oil prices, future exchange rates, future production levels, future capital expenditure levels and future inflation rates. A projection for any one of these five would be subject to considerable uncertainty; all five combined together in revenue forecasts therefore give rise to estimates which involve a very large possible margin for error. Table 2.6 compares five recent estimates of future government revenue from the oil and gas sector.

The striking feature of Table 2.6 is the sharp divergence between the Treasury forecast and the other four forecasters. In 1983 the Treasury forecast a total government revenue figure of £6.1 billion while the other three forecasts are above £9 billion. The explanation of this large difference lies in the specific assumptions which have been made in order to forecast the future tax revenues. The National Institute and Wood Mackenzie figures for total government revenues are close but they are founded on very different assumptions. Table 2.4 shows that

Table 2.6: Estimates of Future Government Oil Revenues

	Treasury	National Institute	Cambridge Economic Policy Group	Phillips and Drew	Wood Mackenzie
£ million (current prices)					
1982	6,200	7,250		8,600	9,181
1983	6,100	9,425		12,000	10,856
1984	7,500			16,600	14,174
1985				19,300	15,261
1986					17,197
—	—	—	—	—	—
1990					13,412
£ million (1980 prices)					
1982	5,100	5,918	6,300	7,000	7,666
1983	4,579	7,075	6,300	8,900	8,385
1984	5,211		7,400	11,200	10,070
1985				11,900	10,000
1986			8,000		10,485
—	—	—	—	—	—
1990			10,300		5,000

Wood Mackenzie's oil output estimates are much higher than the National Institute's, but Wood Mackenzie assume a zero increase in the real price of oil while the National Institute expects positive increase in each year. So the Wood Mackenzie assumption of high oil production and zero real increase in oil prices combines to produce a figure very close to the National Institute's estimate which is founded on more moderate production and price assumption.

It is also possible to justify a wide range of estimates of future capital expenditure. In forming its estimates the Treasury tends to make a uniformly conservative set of assumptions, that is, low output, a low rate of oil price increase, low inflation and fairly high capital expenditure. One of these assumptions is unreasonable taken on its own; together they combine to produce a very low revenue estimate. The other forecasters tend to be much less conservative in at least one or two of their assumptions, although there are no forecasts based on a uniformly optimistic set of assumptions.

The likely path of government revenues beyond the next four or five years is even more uncertain. This is illustrated by the Cambridge Policy Group figures which see tax revenues rising to 1990 while the

Wood Mackenzie prediction for 1990 is only half the 1985 figure (expressed in real terms). The explanation for this discrepancy lies in the oil production assumption; Wood Mackenzie's production figures are based on fields currently producing and under development only. Table 2.4 shows that their production forecast for 1990 is quite low (64 million tonnes). The Cambridge Policy Group's tax revenue estimates are based on the Rowland production estimates which are substantially higher in 1990 (138 million tonnes) as they include a considerable production estimate from fields which are not currently under development. The degree of uncertainty over likely tax revenues beyond 1990 is therefore so enormous as to make detailed projections almost meaningless. While there is certainly a strong possibility that tax revenues will continue to rise through the late 1980s into the 1990s, any long-term economic plan must allow for the possibility that tax revenues will be falling rapidly by 1990.

Notes

1. C. Robinson and J. Morgan, *North Sea Oil in the Future*, Macmillan, 1978.

2. P.R. Odell and K.E. Rosing, *The North Sea Oil Province: an attempt to stimulate its development and exploitation 1969-2029*, Kogan Page.

3. See M.H. Pesaran and L.J. Slater, *Dynamic Regression Theory and Algorithms*, Ellis Horwood, 1980.

3   DEPLETION POLICY

The Governor of the Bank of England said in the 1980 Ashbridge Lecture on the subject of North Sea oil: 'We could raise our living standards by borrowing against it. Or . . . we could raise production of oil to become substantial net exporters of oil in the short term, so as to consume the extra imports that we could buy. In either case we would be living better now, but at the expense of the future. This would, in my view, be wholly misguided'. Similarly the tenor of the 1978 White Paper 'The Challenge of North Sea Oil' is that oil revenues should be invested and not squandered on consumption. But is this a reasonable stance to take; is it really sensible to abstain from consuming the immense wealth represented by the North Sea when the ultimate objective of economic activity is consumption of goods and services? To invest the revenues coming from the North Sea is to transfer consumption into the future at the expense of the present. Is this really a desirable course of action? No one suggests that the output of the coal industry, or any other industry for that matter, should be invested; why is oil different? Part of the answer is that we see oil as something special, a free gift which we do not have to work for. To a large extent this is true; the costs of oil extraction, even in the North Sea, are considerably smaller than the revenues which are generated. The oil industry is therefore immensely profitable, but is this sufficient reason to insist on the investment of the profits rather than their consumption?

In this chapter we discuss the correct management of a natural resource, including the correct rate of depletion of the resource, and the level of increased consumption which the resource should support. The starting place for this discussion will be an account of the main microeconomic factors which theoretically determine these decisions. The economy at large will be taken as given and the development of the natural resource will be assumed to have no effects on the functioning of the rest of the economy. So it is assumed e.g. that oil may be imported if a zero North Sea depletion path is chosen and that rates of return in the economy at large will not be affected by investment resulting from the depletion of the resource.

It quickly becomes apparent that many of the theoretical concepts used are vague or hard to quantify in a real world situation. The second section of the chapter will consider some of the more realistic factors which need to be considered before the theoretical model can be made operational. We end the chapter with an assessment of the government's attitude towards depletion policy.

A Theoretical Approach

It seems reasonable to accept that oil, as a natural resource, is not the property of any one generation. So it follows that it would be wrong for any generation to enjoy the consumption resulting from the resource to the exclusion of all other generations. But this applies equally to future generations consuming at the expense of the present. Both present and future generations are entitled to a share in the total consumption. The problem is to decide how much of a share. It is clearly too simplistic to say simply 'consume it', or 'invest it'; what must be answered is the question 'how much may be consumed?' or 'how much may be invested?' Closely related to this problem is the question of the correct extraction or depletion rate. The formal literature which has grown up around these problems is now quite extensive. Stemming initially from Gray's 1914 paper, and Hotelling's 1931 paper the discussion has spread into many areas of theoretical economics, particularly Growth Theory (see Solow, 1974, and Dasgupta and Heal, 1979, for an overall survey of the subject). This section will confine itself to a micro-perspective on the problem and it will therefore be assumed that the economy will not respond in aggregate to the development of a natural resource. This is not an unreasonable assumption in the case of North Sea oil as it represents a relatively small part of total output, it is obviously not realistic in the case of the OPEC states where oil provides a major share of total production. This assumption will be relaxed later in the chapter and in the subsequent chapter when the general macroeconomic effects of a resource will be explored. The value of the assumption here is that it allows the factors which affect the depletion and investment decisions to be more easily identified. The objective here is to determine the factors which should govern consumption resulting from the resource, investment resulting from the resource, and the actual depletion rate of the resource. There are two distinct approaches to this problem. In the first depletion is seen as being a problem of the allocation of total assets between various investment possibilities, one of which is the natural resource itself. This approach follows Hotelling in setting up some sort of net present value function and maximising it. The basic result yielded is that the rate of increase of the real price of the resource will be equal to an appropriate rate of interest in equilibrium. The problem with this approach, at a formal level, is that it allows no room for consumption. The objective is taken to be to maximise the value of the resource and consumption is completely ignored. If the optimal investment policy were zero depletion the resource would remain untouched and consumption due to the resource would be nil.

The other main approach consists of maximising a social welfare function involving consumption levels over a given time horizon. In its simplest form this approach is no more than the allocation of a given total consumption level over a number of time periods. In this simple form it fails to capture the potential of a natural resource to earn income either in the form of an increase in its real value because of real price increases or in the form of depletion and investment in an alternative, income yielding, asset. This approach can be made more satisfactory, however, by introducing the potential for the resource to earn a real rate of return; this has been done in terms of fairly complex growth models (see Dasgupta and Heal, 1979, Chapter 10). The appendix to this section outlines a small formal model which derives many of the results reached by this type of analysis.

This second approach is the more relevant of the two if the decision-maker is a government or general economic planner. The problem breaks down into two parts. Firstly, how much consumption may be undertaken on the basis of the resource in each time period? This may be consumption of the resource itself or it may be consumption of any product which is financed by borrowing against the natural resource. Secondly, there is the problem of how best to invest; the natural resource represents an asset which is capable of earning a return either by direct increases in its real price or by being sold and invested in some other income-earning asset. The actual rate of depletion will emerge from the solution to these two problems.

This approach can be broken down into two main sections: the structure of the social welfare function and the model of the economy. The specific nature of the economic model does not crucially affect the results and so it will not be considered further in this chapter. The social welfare function, however, contains two essentially arbitrary assumptions which are crucial to the overall results. These are the assumed form of the utility function applied to consumption in each period and the rate of time preference which is applied to these functions. The utility function is simply the formula which measures the value of total consumption in each period; it is usually assumed to have positive but diminishing marginal utility. This means that while more consumption is always desirable an extra unit of consumption will be valued more highly when total consumption is low than when it is high. The extreme form for this function is to assume that it is linear, in which case marginal utility is not diminishing. The rate of time preference applied is a very important determinant of the results. Time preference simply indicates whether there is a preference for consumption in one time period over another. If it is positive it means that we value a given level of consumption in the present more highly than the same level of consumption in the future. A time preference of zero will mean that we do not discriminate between consumption at different points in time.

It is the interaction of these two sets of assumptions which essentially determines the rules for the consumption, investment and depletion decision. The consumption decision will be considered first.

There is no doubt that most of the founders of this 'utilitarian' approach were wholly against the concept of positive time preference and many more recent, eminent, economists have supported their stance. Harrod has called the use of positive time preference 'a polite expression for rapacity and the conquest of reason by passion'. It is not disputed that individuals should, and do, employ positive time preference in making their own consumption decisions. It is, however, argued that the economic planner should be an unbiased arbiter allocating resources between a number of generations. It is, therefore, wrong for such a planner to favour consumption in any given time period. The use of time preference has been justified on the basis of uncertainty about the long-term future and the belief that in a reasonably long time horizon technology can change so drastically as to render current constraints meaningless. It would have been wholly reasonable if stone age man had over-consumed the exhaustible resource of flint deposits, as they have no value today and postponing consumption would have been pointless. Similarly, it is possible that in a hundred years, nuclear fusion or some other development will have made oil less valuable than it is today. However, while this argument is a good one for depleting the resource and storing the purchasing power as some other asset, it is not an argument for consuming it. The consumption decision is not dependent on the future usefulness of the natural resource. The most powerful argument for the use of time preference is the profound uncertainty about the long-term future.

If time preference is not used then conventional utilitarian analysis will always give rise to an upward-sloping consumption path. This is intuitively quite a simple result to understand; if we value future consumption just as much as present consumption then it is always possible to increase total consumption by reducing current consumption and increasing future consumption by the amount of the reduction, plus the interest which it has earned. So it is possible to reduce present consumption by an amount x and raise future consumption by an amount x(1+r) which is greater than x. If the utility function is a simple linear one then it will generally be desirable to reduce current consumption to zero in order to maximise the sum of consumption over time. In other words, all consumption is pushed into the future, and if the time horizon is infinite then actual consumption will be postponed indefinitely. If the utility function used is a more conventional convex one (in particular if the slope approaches infinity as consumption approaches zero) then consumption cannot fall to zero in any period. The overall time path will be a smoothly rising one.

When positive time preference is introduced it will always have the effect of causing more consumption in the early periods. If the rate of time preference is quite small the consumption path may initially slope upwards, reach a peak, and then fall eventually to zero. If the rate of time preference is quite large then consumption may begin at a high level and gradually fall over a time to zero. These two paths can be illustrated diagrammatically in the following way where δ represents the degree of time preference.

Figure 3.1 shows the steadily rising consumption path when $\delta = 0$, the path which rises and then ultimately falls when δ is small and the steadily falling consumption path when δ is large. It is clear then that almost any path can be justified as being optimal if the appropriate assumptions are made about the utility function and the rate of time preference. The 'invest it' argument is derived from a linear utility function and zero time preference; the 'consume it' argument is supported by a high rate of time preference. It may seem that the formal analysis has achieved very little. It does, however, give one important insight, and that is to show what types of consumption paths are possible.

The $\delta = 0$ path makes it clear that there is no reason why consumption need ever fall. That is to say, an often posed question is, what happens when the oil is gone? This is the well-known re-entry problem.

Figure 3.1: Consumption Over Time

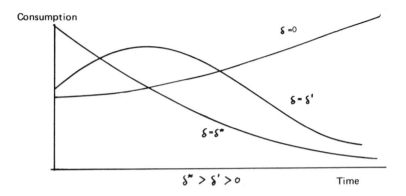

But the $\delta = 0$ path shows a possible future which has no re-entry problem because consumption resulting from the resource never ends.

A slightly different approach to the consumption problem is to ask the question: What is the maximum constant consumption level which may be maintained in every period? This will obviously be a consumption level above the initial level of the $\delta = 0$ path and below the initial level of the $\delta = \delta*$ path. It turns out that in fact this path can be very closely related to the income which the resource earns in each period. This is an intuitively very appealing result; it says simply that if the value of the stock of the natural resource is equal to a given amount, initially, and in each period it earns a real rate of return either in the form of an increase in the real price or a yield from being invested in an alternative asset, then if only this return is consumed the real value of the stock will remain unchanged and so this consumption level may be maintained indefinitely. In effect, this level of consumption is the maximum which may be consumed without transferring consumption from the future to the present.

The formal analysis of correct investment policy for an asset such as a natural resource is less controversial than the consumption analysis as problems of time preference do not arise.

The result which emerges is quite simple and consistent over a wide range of formal models. It is, simply, that the natural resource is only one way to hold wealth within the economy. In order for it to be sensible to hold wealth in the natural resource at all it must yield a real return which is comparable to that given by the other assets in the economy. Clearly oil, or any other natural resource, cannot be productive in the same way as a piece of capital investment, but it can yield a return if the real price of the resource increases. So when the real price of the resource is rising at a rate equal to the return given by the best alternative asset, the planner will be indifferent between investing in the asset and holding the resource as a physical stock in the ground.

The above analysis of the investment decision is often presented as the correct determinant of depletion policy. This is not wholly correct, however. The determination of depletion in any one period should come from a combination of the consumption and the investment decisions. This can most easily be appreciated through a set of examples; first suppose the split of total wealth between the various assets is ideal so that each asset is yielding the same real interest rate. Under this condition the proportion of wealth held in each asset should remain constant; this means that consumption should come from each asset according to the proportion of total wealth which it represents. Under this condition no depletion of the resource should occur for investment purposes, but consumption depletion will occur. The alternative assumption is where the assets are not properly distributed. If it is decided that a major part of the resource should be shifted into some other asset, then depletion should take place at the maximum rate possible until the assets are correctly distributed. It is then irrelevant where consumption comes from. If the natural resource is considered to be the best investment, then it should not be depleted at all. Consumption should then come from the other asset stocks so that non-resource assets are actually reduced by consumption while the real value of the resource grows. When the asset stock is properly distributed, depletion should take place primarily for consumption purposes; when it is not optimally distributed then depletion should primarily be determined by investment considerations.

This consideration of the formal analysis of resource management has brought out a number of points. Firstly, it is not irresponsible to consume rather than invest a natural resource. Secondly, investment and consumption decisions may be reached entirely independently so that it is quite possible to decide to increase consumption and adopt a policy of low depletion. Thirdly, depletion policy should be determined by both investment and consumption decisions, although investment is the important determinant while the asset stock is out of equilibrium.

Applied Resource Management

The formal analysis yields the general principles of resource management. It does not, however, give an unambiguous or simple guide to the practical problems faced by a resource manager as a number of the concepts used in the formal analysis are either highly simplified and abstract (e.g. the rate of interest) or subject to a huge degree of uncertainty (e.g. the path of future oil prices).

The most important determinant of both consumption and investment within the formal model is the real rate of interest. This is not an easy concept for which to find a real world counterpart, partly because there is no such thing as a unique interest rate but rather a whole structure of related rates, and partly because these rates do not remain stable over time. The most cursory glance at the world asset markets shows immediately that there is a very wide range of rates of return offered at any time. The reason that assets with comparatively low rates of return are held is that they have some other advantage. Basically all assets can be assessed in terms of three properties which they possess to varying degrees: their rate of interest, their liquidity and their degree of capital certainty. A highly liquid asset such as a building society deposit might be expected to offer a comparatively low rate of return while a highly illiquid asset such as physical piece of investment might yield a much higher rate of return. Oil is not a very liquid asset, as developing and exploiting an oilfield may involve considerable delay. It is also subject to considerable capital uncertainty, because one can never be sure of the true total value of an oilfield. It is clearly wrong to expect the rate of return on oil to be equal to some market rate of interest. Oil should theoretically hold a place in the whole spectrum of rates of return but just where that place should be is difficult to judge.

Even when the problem of the term structure of interest rates is ignored there is still considerable uncertainty surrounding the real rate of interest. One fundamental distinction which must be made is between the *ex ante* or expected rate of interest and the *ex post* or actual observed rate. The expected rate will generally be fairly stable over time and it is this rate which is important for the decision-maker. The *ex post* or actual rate will often fluctuate quite dramatically when measured over a short period although in the longer term it should be more stable. There is no simple relationship between these two concepts; over a very long run of years the expected rate should not show a consistent difference from the actual rate as this would imply a very inefficient process for expectation formation. At any given time, however, the expected rate may diverge from the actual rate by a considerable amount. The expected rate of interest is not, of course, a unique number: different decision-makers may hold different expectations and so reach different judgements.

The most obvious guide when forming judgements about future rates of interest, or the rate of change of oil prices, must be the past. It is not sufficient simply to say that this was the rate of interest so it will continue. But the past is the best place to begin an assessment of expectations although the future may be radically different from the past.

Table 3.1: The Real Interest Rate Implied by 2½ per cent Consols and the *Financial Times* Industrial Share Index

	Annual % real yield on FT index %	Annual % real yield on 2½% consols %
1956	−7	−0.2
1957	6.6	1.5
1958	−0.5	1.9
1959	42.0	4.3
1960	30.7	4.3
1961	2.0	3.0
1962	−9.8	1.8
1963	12.8	3.6
1964	11.5	2.9
1965	1.8	1.7
1966	.5	3.0
1967	10.1	4.4
1968	30.7	2.8
1969	−10.7	3.7
1970	−16.4	3.0
1971	2.4	0.1
1972	28.5	2.0
1973	−19.7	1.9
1974	−59.0	−1.1
1975	7.4	−12.5
1976	9.2	−2.6
1977	14.8	−4.1
1978	3.6	4.0
1979	−8.9	−2.3
Annual	Implied long-run real rate 1	1

Source: Economic Trends Annual Supplement, 1981 (calculated as an annual rate of return).

Table 3.1 shows the real annual rate of interest for 2½ per cent Consols from 1956-79 and the real annual yield on stock market shares (that is the dividend plus appreciation or depreciation). The stock market gives a much more unstable pattern of real interest rates than the consol market but it is interesting that when the implied long-run rate is calculated for each asset they are remarkably similar. The long-run rate is simply that constant rate which would yield the same value for wealth at the end of the period as the varying actual rates. So, if we invest £100 and three years later it is worth £133 the long-run rate is 10 per cent, even though the actual rate may have been 0 per cent, 0 per cent, 33 per cent. The fact that the two yields are so close over the whole period may well in fact be a coincidence. The rise in inflation since 1974 has certainly affected the pattern of consol yields, while the sudden slump in stock market prices in 1974, due to the oil price rise and the start of the recession, has also distorted this trend. The implied real rate from 1956-73 for the stock market is 5.1 per cent while the consols give an implied rate of 2½ per cent. For the short period 1975-9 the rate for the stock market is 5 per cent, while the rate for consols is 3½ per cent. The government has gone some way towards setting a relevant real interest rate in its 1978 White Paper (Cmnd. 7131). This paper discusses the criteria which should be met by any piece of public investment. The main test is 'the required rate of return' (RRR); this is a minimum real rate of return which should be earned on any investment project. In the July 1981 Economic Progress Report the following statement regarding this test was made:

> The important factor here is the 'opportunity cost' of capital — what the resource would earn in other uses. This is represented by a return which the industries are required to make on new investment. The RRR was set at 5 per cent in real terms — that is after allowing for inflation. The principal elements lying behind this choice were the pre-tax returns which have been achieved by private companies, and the likely return on private investment.

In fact the overall return on capital employed in the nationalised industries has not been anything close to 5 per cent over the last ten years.

It seems that historically the real rate of return from industry generally is close to 5 per cent while the return on consols is approximately 2-3 per cent. It might be reasonable, in view of the current recession, to assess the expected real rate of return from industry at a conservative figure of 2-3 per cent.

Having considered what may be termed a reasonable rate of interest it is now possible to consider what level of consumption could be permanently maintained on the basis of these rates. It may be assumed from the discussion in Chapter 2 that total North Sea oil stocks stand at a figure in the order of 4,000 million tonnes. At 1980 prices this stock would be valued at approximately £450,000 million. This figure greatly overstates the real value of the North Sea stock, however, as the costs of extraction are quite considerable. The Department of Energy has assessed current production costs in the North Sea at an average figure of £35 per tonne, and it expects fields currently under development to yield an average cost figure in the order of £45 per tonne (both in 1980 prices). If the cost of development of the rest of the oil is assumed to rise steadily to a figure of £80 per tonne, this would give a total cost figure for extracting all the oil in the order of £200,000 million. This would put the net value of North Sea oil in the region of £250,000 million. (A more detailed calculation of the value of North Sea oil has been carried out by Scott (1982). He derives a rather lower figure as he assumes a much higher escalation in costs.) Table 3.2 shows the maximum permanent consumption level which would result from the choice of various real rates of interest.

Table 3.2: The Real Rate of Interest and Maximum Permanent Consumption

Interest rate (%)	Permanent consumption £m
½	1,250
1	2,500
2	4,900
2½	6,000
3	7,300

On the basis of this calculation, if the real rate of interest is taken at 2 per cent then consumption of the order of £5 billion can be maintained indefinitely. In view of the fact that the total value of oil production in 1980 was only £8.8 billion and that a large proportion of this was devoted to paying back early capital expenditure, so that total government receipts were only £3.8 billion, suggestions that all this revenue should have been invested rather than consumed seem to be ill-founded. Indeed, as Table 3.3 shows, the real return which has been earned by oil over the last 10 years has been much higher than this 2 per cent.

Table 3.3: Real and Nominal Oil Prices and the Real Return on Oil

	Nominal oil price £/tonne	Real price of oil £1980/tonne	Real rate of return on oil %
70	7.4	27.03	—
71	9.29	31.01	14.7
72	9.37	29.17	−5.9
73	12.1	34.5	18.2
74	36.1	88.79	157.3
75	38.05	75.33	−15.1
76	51.1	86.8	15.2
77	57.7	84.62	−2.5
78	51.4	69.6	−17.7
79	66.9	79.8	14.6
80	104.0	104.0	30.3

Source: National Institute Economic Review, *Economic Trends*.

The investment decision is perhaps much more difficult to reach than the consumption decision. Consumption entails only formulating an idea of the expected long-run real rate of interest and using this to decide what consumption levels may be sustained. The investment decision, however, depends crucially on the relationship between the expected real rate of interest and the expected real rate of oil price increases. Predicting future oil prices, even in the short term, is a process which is enormously uncertain and when this needs to be done on a fairly long-term basis the uncertainty becomes unmanageable. Table 3.3 shows the movement of oil prices through the 1970s.

The long-run annual rate of return earned by oil over this period is 14 per cent; not surprisingly, therefore, oil is seen as a very good investment in comparison with either the stock market generally or financial assets such as consols. However, it is quite unrealistic to assume that this rate of real price increase can continue indefinitely. If this were to happen the real price of oil, at 1980 prices, in the year 2000 would be £1,500 per tonne. It may well be asked why such a price is beyond the bounds of possibility, given that in 1970 no one would have predicted the subsequent oil price rises. The answer to this lies in the possible substitutes which exist for oil. In the short term there are, for many purposes, simply no substitutes, but in the longer term there are various possibilities. One is simply that other fuels may be substituted for oil by changing the type of technology being used, so that oil-fired boilers may be changed to coal, gas or electric, for example. This option is important, but there are many applications for which it is simply not suitable, such as aviation, or road transport. Another alternative is that liquid fuels can be manufactured from coal; the National Coal Board is currently developing two separate industrial plants to carry this out and there is a wide range of other synthetic fuel processes being developed. Petrol from coal is obviously more expensive to produce than petrol from oil, at cost price, but if the price of oil goes high enough the coal-based production process will become financially viable. So the price of liquid fuels from coal is likely in the long run to impose an upper limit on the real price of oil. The actual cost of coal produced liquid fuels is, of course, subject to uncertainty, but there is little doubt that if the price of oil were to reach £150 per tonne, in 1980 prices, then there would be a strong incentive to begin liquid coal fuel production. Of course, in the medium term the price of oil could go well above this figure as it would take many years to introduce the large-scale conversion of coal, but this process must set an upper limit to oil prices in the long run.

It seems reasonable to assume, therefore, that the real price of oil cannot continue to rise over the long-term future, say the next 20 years, at a rate anything like that of the past ten years. In the short term, the next five years or less, oil may still be a preferable investment, but annual production rates limit the speed with which oil may be converted into other assets and so it is the longer term rate of price increase which is relevant to the investment decision. In terms of the other measures which might normally be used to assess an investment, liquidity and capital certainty, oil seems only a moderately good investment. It enjoys a fair degree of capital certainty, its real value is unlikely to fall substantially over the long run, but it is an asset which is very hard to realise quickly: it is highly illiquid.

It is possible, however, that oil may be considered to yield a more competitive rate of return if the concept of a rate of return is widened. Oil is not simply an asset, it is also an important part of the general industrial process of the economy. Much of industry simply cannot function in its present form without oil. It is valid to ignore this special quality of oil and to consider it merely as an asset only if we are completely certain that we can obtain the oil required by the economy from other countries. This assumption was made in the first section. If oil supplies generally are uncertain, however, then the domestic oil stock may be considered to yield a fairly considerable return in terms of increased security.

It is clear that one of the deciding factors in decision-making is how highly we value the security aspect of North Sea oil. This factor cannot be considered apart from a world context. If energy is plentiful and freely available, the security factor will have little value; if energy is either scarce or located in politically unsettled areas the security factor may become an overwhelming consideration. Similarly, if there were to be a major shortfall in world oil supplies, the international pressure on the UK to increase oil exports would be considerable.

It is important, therefore, to know what the overall world energy prospect actually is. This is particularly true in view of the general acceptance of the notion of an energy crisis. The popular idea that oil will run out in approximately 30 years at present production rates is a prime example of the misinterpretation of statistics. In fact throughout almost all of this century proven oil reserves have been sufficient to last about 30 years. All this means is that oil companies like to maintain this level of discovered stocks in the ground. It says nothing about how much oil actually exists or how long it is likely to last.

When considering this security factor it is necessary to look at both the world energy situation and, more particularly, the state of the world's oil resources. These two facets of the problem will be approached separately.

World Energy Prospects

In order to put world energy supplies and resources in perspective it is first necessary to consider the world's overall energy needs. It is important to understand that just as the figures for any one country are subject to a degree of uncertainty, so figures for the world as a whole are even more uncertain. The data presented here is therefore intended to represent orders of magnitude rather than exact figures. Table 3.4 gives some figures for the rate of world energy consumption in 1980.

Oil is currently supplying some 43 per cent of the world's primary energy needs. Coal and natural gas are supplying 29 per cent and 18 per cent, respectively, while the contributions of hydroelectricity and nuclear power are very small indeed. The figures presented here for energy reserves and resources are based on those of the 1980 world energy conference. These figures are constructed on a country-by-country basis. Proved recoverable reserves are defined as those discovered reserves which are currently commercially exploitable. The

Table 3.4: Estimated Annual Consumption of Primary World Energy Supplies

Fuel	Quantity (million tonnes of oil equivalent)
Oil	3001.4
Natural gas	1278.3
Coal	2020.9
Hydroelectricity	414.6
Nuclear	167.4
Total	6882.6

Source: BP Statistical Review of the World Oil Industry, 1980.

difficulty with the concept of the total available resource is that it is impossible to conceive of a time when all of a resource will have been extracted. Some coal or oil will always remain in the ground simply because it is so sparsely deposited that it would cost more to remove it from the ground than it would be worth. So the concept of a total resource needs to be bounded by a constraint on the lowest acceptable quality of the resource being considered. Different researches may impose different limits and so the estimate of additional resources is subject to a wide degree of uncertainty. The estimated additional resources are defined as the total quantity of the resource available for exploitation. This ignores normal commercial considerations but excludes deposits of such low quality that they are unlikely ever to be exploited. Table 3.5 presents figures for the world's energy resources.

Table 3.5: Total World Conventional Energy Resources

Type	Proved recoverable reserves (mill. toe)	Estimated additional resources (mill. toe)
Coal	453,790	6,612,600
Oil	88,900	212,000
Gas	61,700	159,870
Uranium		
Used in current generation nuclear reactors	43,000	58,100
Used in fast breeder reactors	1,500,000	2,932,000

Source: World Energy Conference Survey of Energy Resources, 1980.

There is no simple division between the two categories presented above. In the short term it is true that only the proven recoverable reserves can be considered in planning, but over time a large proportion of the additional resources can be moved into the proven category simply by carrying out exploration. So at a rate of consumption of 3.0 billion tonnes of oil a year the proven recoverable oil reserves will last for only 29 years but the additional resources could last for a further 50 or 60 years. From the standpoint of energy generally current gas supplies could be maintained for 150-200 years, and coal production could be maintained almost indefinitely (even the proven recoverable reserves could last almost 300 years). There is no energy crisis as such; conventional energy sources are sufficient to last almost indefinitely. There is, however, a relative oil shortage; given our present consumption patterns oil will become scarce long before the other fuels.

Does this apparent abundance of energy imply that the North Sea oil stock has little value as a hedge against an energy crisis? This is not entirely the case; it ignores the fact that energy, and oil in particular, is not evenly spread throughout the world. If energy were distributed evenly according to consumption patterns, then the North Sea would have little special value. However, some areas of the globe will exhaust their indigenous oil supplies long before others, and it is this inequality which gives the United Kingdom oil a unique value.

The International Oil Market

On a worldwide basis oil is unlikely to become scarce within the next 50 or 60 years. The cost of producing oil will, of course, rise as exploitation takes place in less favourable regions, although this is not particularly important as world oil prices are already far above production costs in most onshore production sites. On a regional basis, however, the situation is very different.

There are approximately 600 sediment basins around the world, that is sites where deposits may accumulate. The known hydrocarbon deposits are located in 160 of these basins and some further 240 basins have been explored without yielding any economic discoveries (although future exploration in these areas may still be profitable). Some 200 basins have not yet been explored at all, although these generally lie in particularly hostile regions. Table 3.6 gives details of the general location of oil reserves and resources as well as cumulative production so far.

Table 3.6: World Oil Reserves and Resources (million tonnes)

Region	Total production before 1980	Proved recoverable reserves 1.1.81	Additional recoverable resources	Total oil originally in place
Africa	4,046	7,400	34,000	45,446
North America	18.083	5,000	24,000	47,083
Latin America	7,336	9,700	12,000	29,036
Far East	1,845	2,700	12,000	16,545
Middle East	15,607	49,200	52,000	116,807
Western Europe	686	3,100	10,000	13,786
USSR, China Eastern Europe	6,257	11,800	64,000	82,057
Antarctica	—	—	4,000	4,000
Total	53,860	88,900	212,000	354,760

Source: World Energy Conference 1980; BP Statistical Review of the World Oil Industry, 1980.

Africa has now used 9 per cent of its original total oil stock, North America 38 per cent, Latin America 25 per cent, the Far East 11 per cent, the Middle East 13 per cent, Western Europe 5 per cent, the communist countries 8 per cent and the Arctic stock remains unused. The regional inequality is already beginning to emerge; North America has produced 33 per cent of total production although it only possessed 13 per cent of the world's original oil resources. This inequality becomes even more marked, however, when the reserve production and the reserve consumption ratios are calculated. These figures are presented in Table 3.7.

The variation between regions is enormous, with the Middle East able to sustain its level of consumption for 600 years, while North America and Western Europe would both exhaust their reserves in under 6 years. The picture which emerges is not one of a division between the communist and non-communist countries but a division between the oil owners and the oil users. The communist block is largely self-sufficient in oil and can be ignored in terms of the major split between the Middle East, which has oil, and Western Europe, North America and the Far East, which consume oil. Of these three regions only North America originally had a sizeable oil resource in global terms, and this resource has already been exploited to a greater extent than any other region. This is illustrated by the fact that by

Table 3.7: The Reserve Production Ratio and the Reserve Consumption Ratio

Region	Consumption 1980 (million tonnes)	Production 1980 (million tonnes)	Reserve Production Years	Reserve Consumption Years
Africa	71.9	296.9	25.0	103.0
North America	878.9	508.2	9.8	5.6
Latin America	222.1	296.6	32.7	43.6
Far East	399.0	124.7	21.6	6.7
Middle East	82.0	927.0	53.1	600.0
W. Europe	682.5	126.0	24.6	4.5
USSR, China E. Europe	626.6	727.5	16.2	18.8

1976 a total of 3,300,000 wells, of all types, had been drilled throughout the world and over 2,500,000 of these had been drilled in North America. Less than 12,000 wells had been sunk in the Middle East, the world's largest oil resource. This high level of exploitation in North America also shows itself in the low average production of a well. In Saudi Arabia average production is in excess of 350,000 tonnes per well per year; in Iran it is in excess of 500,000; and in the UK it is around 475,000. In the USA average production per well is around 850 tonnes per year.

It is within the context of the heavy dependence of the Western industrialised countries on Middle East oil that the UK's North Sea oil finds its special place and value. There is no world energy shortage as such, nor is there even an imminent world oil shortage. But if the flow of exports of oil from the Middle East were to suffer a major reduction, either through war or direct political action, then the Western countries would face a major economic upheaval. UK oil production could do relatively little to help the Western world in such a situation, but it could do a great deal to protect the UK economy and in so doing it would relieve at least part of the burden on the other countries.

A Depletion Policy

It has been argued that a reasonable expected annual real interest rate might be of the order of 2-3 per cent. The real price of oil is unlikely to rise beyond a level of £150 per tonne (1980 prices), which over 20 years would imply a 2 per cent per annum real rate of price increase and that there must be a distinct possibility of even this rate of increase not being achieved in view of the relatively abundant world oil supply situation. On a simple comparison of these two expected rates of return it seems that oil is a relatively poor long-term investment and that a policy of high depletion rates should be introduced so that the wealth represented by the North Sea be converted into other assets. However, this result ignores the security value of having an indigenous stock of oil. The difference between the two rates given above is so small that the security aspects of oil would not have to be given a very high weight in order for it to tip the balance in favour of low depletion.

There is, however, a further complication to the security factor, which is that the oil stock in the ground does not provide security of oil supplies in itself. The necessary factor which provides security is the actual potential for production at a level which is at least close to self-sufficiency. So if we judged the oil stock to yield an additional return of 2 per cent per year for its security value, this would make it a sound long-term investment and indicate a zero depletion policy. But in practice it is impossible to maintain a potential for oil production for any length of time without actual oil production. This is simply because no private company would engage in the development expenditure necessary for an offshore oil field without the right to produce oil. Even a national government would find such development and maintenance costs hard to finance over a long period without oil production. In order to maintain potential production levels actual oil depletion must proceed at a level close to the level of likely future domestic oil demand. Production beyond this level would be undesirable as it would reduce the number of years of potential oil production and thereby reduce the security value of the oil.

Table 1.8 in Chapter 1 gave figures for recent domestic oil consumption levels. These figures have fallen dramatically from close to 100 million tonnes in the early 1970s to 74.4 million tonnes in 1981. The recent very sharp falls in 1980 and 1981 must be largely due to the general economic recession. If the economy eventually moves out of this recession the demand for oil would certainly increase enormously, perhaps to between 90 and 100 million tonnes per year. This possible demand range should therefore be the aim of a depletion policy which is being determined by the security aspects of oil. Any depletion beyond this range would be undesirable.

The Government's Attitude Towards a Depletion Policy

Throughout most of the 1970s the government's main aim was to achieve a self-sufficient level of oil production as quickly as possible. Statutory powers had been taken which allowed the government a wide degree of direct control over the rate of oil depletion. The 1975 Petroleum and Submarine Pipeline Act gave the Secretary of State for Energy very considerable powers to regulate oil depletion in either the 'national interest' or a 'national emergency'. These powers were widened even further by the 1976 Energy Act. In addition the licensing system allows the government to control long-term depletion through the rate of exploration and discovery. In order that these powers should not delay the development of the oil sector, the government issued a number of assurances designed to reassure the oil companies. In December 1974 Mr E. Varley, as Secretary of State for Energy, gave the following assurances:

1) For discoveries prior to 1975 no development delays would be imposed and no production controls would be made until 1982 or four years after the start of production.
2) For future discoveries made under licensing rounds 1-4 no production cuts would be made until 150 per cent of the investment had been recovered.
3) Any production delays would be with full consultation.

The major restriction on the government's ability to control depletion was contained in the first point. As Table 1.3, Chapter 1, shows, the vast majority of proven oil reserves had been discovered by 1974 and therefore were protected from any government intervention until 1982. This, in effect, protected the oil companies during the build-up of production to self-sufficiency without limiting the government's abilities to control depletion beyond this date.

Self-sufficiency was achieved in 1980 and oil production was expected to continue to rise. The government no longer had any strong motive to increase the rate of depletion. The arguments given above suggest that the crucial factor in determining depletion policy is the weight which the government gives to the security value of North Sea oil. It is hard to assess the attitude of the government on such a point but the indications are that a high value is attached to the security factor of oil. In evidence given (6 May 1981) to the Parliamentary Select Committee on Energy, the Under-Secretary made the following statements with regard to the Secretary of State for Energy's views on depletion policy: 'I think one of the subjects, or the main subject he attached importance to was security of supply for the nation as a whole'. And later: 'This is a major consideration which led him to adopt a depletion policy and to try and roll forward some of the hump into the 1990s'.

The following statement was made by the Secretary of State for Energy in July 1980, and was considered of sufficient importance to be reproduced in the 1981 Brown Book.

Recent events underline the fragilities of the world energy scene. The Government believes that on strategic and security of supply grounds it is in the national interest to prolong high levels of UKCS production to the end of the century. This requires action to increase exploration, which we have already taken, and to defer some oil production from the 1980s. Such action accords fully with the recommendations to maximise indigenous hydrocarbon production on a long-term basis and with our other international commitments, including net exports of 5 million tonnes in 1985 as agreed in the Community and the International Energy Agency.

There are of course major uncertainties about future levels of North Sea production and UK consumption. There can therefore be no rigid plan. We shall continue close supervision over reservoir performance at existing fields and scrutinise new applications for field developments to ensure good oil field practice consistent with optimum oil and gas recovery in the national interest. We shall also continue to take decisions on a case-by-case basis, but giving greater emphasis to the need to limit the sharpness of the peak in production. We shall, of course, honour the assurance given by the Rt. Hon. Member for Chesterfield [Eric Varley] on 6 December on the basis of which heavy investment has been undertaken by the oil companies.

In particular the Government will consider delaying the development of fields discovered after the end of 1975, which are not covered by the assurances given by the Rt. Hon. Member for Chesterfield. The Government will also continue to tighten control on gas flaring. The Government has taken no decisions on whether to have production cutbacks which, under the assurances given by the previous administration, cannot be made before 1982.

I believe that this flexible approach is the right one and takes account of both the needs of those involved in the difficult business of oil production and, more important, the long-term national interest.

This intention to roll forward the hump in oil production had already begun with the two-year delay in the start-up of the Clyde field. Production in this field would probably have started in 1985, but due to the direct intervention of the Department of Energy, production was postponed to 1987.

More recently government ministers have suggested that there is no need for an active depletion policy. In October 1981, Mr Nigel Lawson made a speech to the Norwegian Petroleum Institute in which he argued that uncertainties about oil demand, supply and future prices were so great that any depletion policy was subject to enormous uncertainty. He also questioned the advisability of government intervention in the oil market.

Equally doubtful is the ability of Governments to weigh all the uncertainties more accurately and come to a better conclusion about the future than the market. After all, the market has the twin advantages of a diversity of view, and a hard financial test of performance.

This is not, however, a total departure from the objectives set out earlier. The change has come about because of two related factors; firstly, in the government's overall economic strategy there is a strong need to reduce the public sector borrowing requirement and high oil production and tax revenues will help to achieve this. Secondly, over the last few years the Department of Energy's estimate for total oil production during the middle 1980s has fallen sharply. In 1979 the forecast for 1983 was 115-140 million tonnes; the current forecast for 1983 is 90-115 million tonnes. So the need for an active depletion policy has been reduced. Indeed in his Norwegian speech Mr Lawson gave the following account of the likely depletion path:

This year was the first in which oil production from the UK Continental Shelf exceeded Britain's domestic requirements. This surplus seemed likely to last into the early 1990s and could amount to between two-and-a-half and three years consumption.

This figure for net exports would seem to be consistent with an average production level in the region of 110 million tonnes a year.

The emphasis of government thinking during the first half of 1982 has increasingly stressed the need to maintain high levels of exploration and development work in order to secure adequate oil production rates during the 1990s. It has been recognised that a strict depletion control now could damage the long-term rate of development and possibly reduce the security of oil supplies towards the end of the century. The House of Commons Select Committee on Energy published a report on depletion policy in May 1982. This report came out firmly against an active depletion policy and recommended that the government should restrict itself to monitoring production levels.

On 7 June 1982 the Energy Secretary (Mr Nigel Lawson) told the House of Commons that no depletion controls would be imposed on fields in production or currently under development before 1984. He stressed, however, that the government retained the right to delay the development of new fields as well as to restrict the development of oil-fields which might involve the loss of associate gas production.

It would seem, therefore, that the official attitude towards a depletion policy has come full circle. Originally no active depletion policy was pursued so that high long-term production levels would be assured. In the late 1970s the expected production levels seemed so high that an active depletion policy seemed to be necessary in order to maintain self-sufficiency throughout the 1980s. By the early 1980s it had become clear that these very high production forecasts had been over-optimistic and the emphasis once more returned to ensuring long-term supplies through encouraging exploration and development.

Conclusion

The theoretical analysis of resource management yields general guidelines but very little in the way of practical concrete advice. The many qualifications which must be made to the theoretical model in terms of its concepts and the evaluation of uncertain future variables, such as the price of oil, prevent any simple and direct conclusions from being reached. It is our belief that the security value of oil is highly important and so the potential for self-sufficiency must be maintained for as long as possible. As oil companies will not engage in exploration and development without the incentive of current production, it is best to allow production at about current levels which are close to self-sufficiency. To go far beyond this level is to sacrifice future self-sufficiency, while to go below it will mean large short-term economic repercussions (in terms of lost government revenues and balance of payments effects) as well as a probable decline in actual productive potential so that full self-sufficiency cannot be quickly achieved in a time of crisis.

It would seem that, given likely oil production and consumption levels, in fact depletion will occur at a somewhat higher level than self-sufficiency. The argument for allowing this is that the damage to future exploration and development rates of actually restricting current oil production would more than offset any possible gains in terms of long-term self-sufficiency. This claim cannot be easily assessed but the determining factor must be the actual level of net exports. Very large net exports would undoubtedly create the need for an active depletion policy while moderate net exports might not. The uncertainty over future production levels therefore requires that the power to impose a strong depletion policy be retained. The current production estimate from the Department of Energy for oil production in 1985 is 95-130 million tonnes. If domestic demand were to return to a level of 90 million tonnes a year then this production range could imply either only a modest 5 million tonnes of net exports or a very sizeable 40 million tonnes of net exports. The larger figure would certainly require government intervention, the lower would not. This emphasises the need not only to bear in mind a target depletion path but also to retain the flexibility of control which can respond to this order of uncertainty.

Part II

OIL AND MANUFACTURING INDUSTRY

4 THE EFFECT OF OIL EXPLOITATION ON THE INDUSTRIAL SECTOR

In the early seventies the prospect of North Sea oil was seen as a major boon for the whole economy. It was generally felt that the balance of payments effects would be such as to remove this factor as a constraint on economic growth. It was further supposed that if the resources generated by the oil sector were to be directed into the industrial sector, a major revival of British industrial prospects could result. More recently however, it has been suggested that the North Sea sector could well operate to the general detriment of the industrial base of this country, that the development of North Sea oil would certainly give rise to a relative decline in the level of manufacturing output and that this decline might, in fact, be an absolute one. The arguments put forward in this debate are so fundamental to the problems of planning a long-term economic strategy for North Sea oil that they will be summarised and discussed in this chapter. The debate is in fact far more wide reaching than its UK context, there are basic theoretical propositions which apply generally to all countries.

The suggestion that the development of a natural resource implies a necessary relative decline in manufacturing industry was made in a paper by R.G. Gregory in 1976, although the idea was in many ways only a special case of an argument made by Meade and Russell in 1957. Gregory outlined the structural shifts which were likely to occur in the Australian economy as a result of the development of a large-scale mining sector. A simple model was constructed which stressed the effect of relative domestic prices on the supply of exports and the demand for imports. The prices considered were the prices of internationally traded goods, imports and exports, relative to the price of non-traded goods. A small country assumption was made, i.e., that world prices were unaffected by the movements of the Australian economy. Gregory derived a simple diagram (Figure 4.1) which had the price of traded goods relative to non-traded goods on the vertical axis and the quantity of imports and exports on the horizontal axis. The lines marked X represent the export supply curve; a higher price in tradeables relative to non-tradeables brings forth a large quantity of tradeable goods for export. The lines marked M represent the demand

Figure 4.1: The Demand for Imports and the Supply of Exports

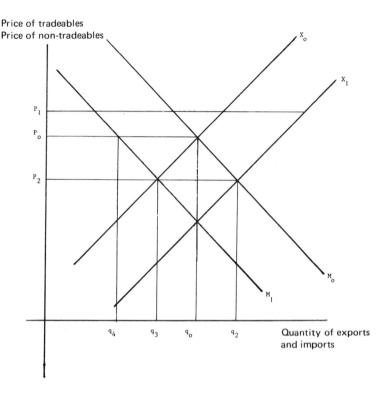

curve for imports, a higher relative price of tradeables produces a
smaller demand for imports. Equilibrium in the balance of trade implies
a relative price P_0 and quantity q_0 of imports and exports. If this
situation were not pertaining, say P_1, where exports are greater than
imports then one of two things would happen: either the exchange rate
would change to change the price of tradeables ($P_T = P_W/e$ where
P_T is the domestic price of tradeables, P_W the world price of tradeables
and e the exchange rate). In the case of P_1 with a surplus the exchange
rate would rise and so P_T would fall. Or domestic inflation would
change the price of non-tradeables while the price of tradeables was
held constant at world prices. In the case P_1, domestic non-tradeable
prices would increase until the ratio P_0 was reached. The case which

seems more important and which Gregory himself focuses on is when the exchange rate adjusts.

So all that has been demonstrated is that the exchange rate will move to bring about a balance in trade. Now it will be supposed that a new sector appears in the economy, such as the exploitation of a newly discovered natural resource. First let us suppose that nothing changes on the import side but that the natural resource is wholly exported. The original import demand curve remains stable (M_0) but the total export supply curve will move from X_0 to X_1. X_0 will represent the supply of the old export items and the horizontal distance X_1-X_0 will represent the new natural resource exports. Clearly the new equilibrium will be at the point P_2q_2 where both total exports and imports are larger and the relative price ratio is smaller (and the exchange rate is higher). Exports other than the new resource will, however, have fallen; in the old equation exports were q_0 now, with the natural resource and a price ratio P_2, non-resource exports will be only q_3. So there has been a rise in the exchange rate causing a decline in the level of non-resource exports and a rise in the level of imports.

The alternative extreme assumption is that none of the natural resource is exported but it is all substituted for imports. In this case the demand curve for imports will fall from M_0 to M_1 and the new equilibrium will be at P_2q_3. The same price ratio will result and the same non-resource trading pattern, exports fall from q_0 to q_3 and non-resource imports rise from q_4 to q_3. Clearly the size of the relative adjustment will depend on the elasticities of the demand and supply curves and the quantity of the natural resource. It is, however, unambiguous that imports must rise (or remain unchanged) and exports must fall (or remain unchanged) and therefore demand for the goods of the tradeable sector must fall.

Gregory does qualify this point by making it clear that this is merely a comparative static exercise and that if a growing economy is being considered the decline in the tradeable sector need only be a relative decline and not an absolute one. Gregory further notes that if by direct government intervention or by rigidities in the market system the relative price ratio does not fall then a balance of trade surplus will result and large quantities of foreign assets will be built up. However, no benefit of the natural resource will be felt within the economy until this adjustment of price ratios is made.

Gregory goes on to consider the likely short-term disequilibrium effects of the structural shift in the economy. These are likely to take two forms: either calls for a devaluation or requests for specific assistance to the tradeable goods sector (i.e. subsidies). Devaluation cannot have any lasting effect on the tradeable sector because if the price of tradeable goods is maintained by a low exchange rate this will merely force the price adjustment to take place in domestic non-tradeable prices through inflation. Even if the old price rate P_0 could be maintained this would not be desirable as it would completely abolish the benefits of the natural resource to the domestic economy. Subsidies similarly cannot protect the tradeable sector as they will simply result in larger balance of trade surpluses and even further revaluation of the exchange rate.

There is in effect no way to escape the relative decline in the tradeable sector other than investing the resource revenues overseas. Complete protection involves investing all the resource revenues overseas and effectively precludes the domestic sector from enjoying any benefits at all.

This general analysis holds good for the introduction of any new sector into the balance of trade. The development of any natural resource which will be either exported or will replace existing imports will necessarily lead to a relative decline in the domestic production of tradeable goods. This conclusion, that the manufacturing sector of the economy will suffer a relative decline as a result of exploiting a natural resource, is often seen as an undesirable one. A decline in exports and manufacturing output is seen to be nécessarily a bad thing. This is not, however, a reasonable view to take. Exports are not, of themselves, desirable things. The reason why a country exports goods is solely so as to pay for its imports. Exporting means that domestically produced goods are sent overseas for other consumers to enjoy. If, therefore, any given level of imports can be maintained (or even increased) with a reduced level of exports then this situation must be seen as desirable. Of course if we are not exporting so much there must be a relative shift in domestic production patterns away from the export sector. This is a wholly desirable shift in the basic structure of the economy. It is, of course, true that this shift may involve unemployment as labour moves from the tradeable to the non-tradeable sector but the object of an economy is not to produce employment but to produce consumption goods. The effect of the natural resource will be to increase total consumption so if the correct pattern of income distribution is adopted every member of the economy could experience an increase in consumption. In the absence of direct government intervention it is highly unlikely that this pattern of income distribution would be achieved. The likely pattern of income changes will be that the owners of the natural resources and those employed in the non-tradeable sector will experience income gains while those people employed in the tradeable sector will experience income losses (particularly if they become unemployed). A legitimate form of government intervention would therefore be to redistribute some of the income gains. But this should be done directly by unemployment benefits, aids in locating new jobs, retraining schemes, etc, rather than by supporting the tradeable sector itself. The structural shift in the economy must take place if the natural resource is to be a positive benefit to society. All that should be done is to either slow and prolong the period of structural change by investing overseas or to protect the members of society who suffer a real loss of income by direct government aid.

This analysis was applied to the UK and North Sea oil by P.J. Forsyth and J.A. Kay (1980). The theoretical stance taken by Forsyth and Kay was almost identical to that of Gregory, although their presentation differed considerably. Forsyth and Kay, instead of adopting a formal model, compared the structure of the UK economy in 1976 (a non-oil economy) with an economy which had undergone the structural adjustments to the introduction of North Sea oil.

This comparison was carried out in a highly stylised fashion. The economy was divided into five sectors: primary production, manufacturing, construction, distribution and services and public administration. Actual levels of production, imports, exports and consumption for each of these five sectors was then presented for the UK economy in 1976. 1976 was chosen as the base year on the assumption that it represented the UK economy without any domestic oil resources, despite the fact that North Sea oil production had begun in 1975 and that total oil production in 1976 was over 12 million tonnes. The figure for primary production was then increased by £10 billion to simulate the introduction of North Sea oil. The other sectors of the economy were then adjusted for this change according to a fairly simple and arbitrary set of assumptions. The overall pattern of changes showed that while the value of total consumption rose by an amount exactly equal to the value of oil production, the pattern of output changes was very uneven. There was the obvious very large increase in primary production, smaller increases in construction, distribution and services and public administration and an actual decline in manufacturing.

This example forms an almost perfect illustration of Gregory's thesis. The increase in oil output causes an unambiguous rise in total consumption; overall production also rises considerably, but there is an important degree of structural change within the productive sector of the economy. This example allows no growth in the economy except that due directly to the oil sector — this is therefore a purely comparative static exercise. That is, it illustrates what would happen if the only change made in the economy were the introduction of the oil sector. If the economy were allowed to grow at the same time as the oil is introduced then the decline in the manufacturing sector might only be a relative one rather than an absolute one. If the economy were to grow by 6 per cent over the period of the oil introduction this would be sufficient in the Kay example to offset the decline in the manufacturing sector. If the introduction of oil took four years this rate of growth would be equivalent to an annual rate of less than 1.5 per cent.

Forsyth and Kay go on from this basic analysis to show that the gains from oil are in fact even greater than this might suggest. In their example total consumption goes up by £10,000 million, the value of oil output, but the changes in export and import levels are brought about by a rise in the exchange rate. This means that the nominal values for imports after the introduction of oil will, in fact, represent a much larger physical quantity of goods imported than would be implied by a fixed exchange rate. In effect, if the exchange rate had originally been $1.50 and we imported £10,000 million worth of cars which represented 2 million cars at £5,000 each and the exchange rate then rose to $3.00, a constant value of imports (£10,000 million) would now represent 4 million cars at £2,500 each. The gain to consumers at the old prices would be an additional £10,000 million worth of cars. It is quite conceivable, therefore, that if the exchange rate rises substantially the gains in terms of more import goods for the same money can actually be larger than the direct gain from oil. Exactly the same situation exists with respect to exports, when the exchange rate rises the same total value will represent a much lower quantity of actual physical exports. If the oil produced a 10 per cent increase in the exchange rate, Forsyth and Kay estimate the indirect gains in this form to be worth £3-4 billion (1980).

Much of Forsyth and Kay's argument can be seen as being completely non-controversial; they are simply following Gregory's analysis of the economic effects of exploiting a natural resource. They do, however, go beyond the strict bounds of his analysis by making general suggestions for an investment policy and by asserting, at least in their numerical example, that the decline in the UK manufacturing sector will be an absolute one rather than a relative one. With respect to their investment suggestions, they argue that investment decisions should be made on the basis of respective rates of return in the domestic sector, the foreign sector, and the oil sector itself. They argue against direct domestic investment on the grounds that they believe the rate of return in the domestic sector will be lower than in the other two sectors. This, in effect, means that the domestic manufacturing sector is to be left to decline at its own rate without any direct government intervention. These points gave rise to a widespread controversy among economists after the publication of the original Forsyth and Kay article.

One of the main areas of confusion which obscured the early debate was whether the analysis implied a necessary absolute decline in the output of manufacturing industry. In their example Forsyth and Kay showed an absolute decline and at a number of points in their paper they seemed to suggest that this absolute decline must occur. The original Gregory arguments presented earlier showed, however, that only a relative decline was inevitable. If Forsyth and Kay's example had been extended to allow sufficient general growth the absolute decline in manufacturing could have been prevented. In a letter in the *Guardian* (8 December 1980) Kay made it clear that his analysis did not necessarily imply an absolute decline in manufacturing but only a relative one. He went on to argue that the growth of the UK economy during the late seventies was so low as to have made an absolute decline in manufacturing inevitable.

The most controversial policy recommendation of Forsyth and Kay was their suggestion that the oil revenues should be largely invested overseas. This recommendation does not follow directly from the formal Gregory model, but it is based largely on the rules for appraising any investment policy. That is that investment should flow into whichever sector yields the highest rate of return. Because they saw the domestic manufacturing sector to be a declining one, they argued that its expected future rate of return would be relatively low and so investment should generally flow elsewhere. This argument is, however, very limited by the assumptions of their model. If the revenue from the North Sea were to be used via fiscal policy to expand the level of economic growth it would be quite conceivable that the absolute growth rate of manufacturing industry could be increased by the presence of oil, even though its relative share of output might fall. This would imply that the rate of return in manufacturing industry certainly does not have to fall as oil production increases. An additional point in favour of domestic investment is that a government may value the secondary effects of domestic investment quite highly; these effects would be such things as increased domestic employment, reduced social security and unemployment payments and any further 'multiplier' effects on domestic production which might result. So that even if, quite apart from the effect of oil depletion, the domestic rate of return were less than the foreign rate, a government still might prefer domestic investment.

One of the most sustained and powerful criticisms of Forsyth and Kay has come from the Bank of England. Part of the intuitive appeal of Forsyth and Kay is that their argument seems to be so well supported in terms of real world events. The late 1970s and early 1980s did see a large increase in the exchange rate and a fall in manufacturing production levels. The Bank has argued, however, that this is a pure coincidence and that in its proper historical context the Forsyth and Kay effects cannot explain these events. This argument was first put forward in the 1980 Ashridge lecture by the Governor of the Bank of England. The heart of his criticism is that while he completely accepts their formal analysis he denies the validity of taking 1976 as a suitable base year. The UK economy in 1976, according to the Governor, was to all intents and purposes an oil economy. It is true that little oil was being produced but a large balance of payments deficit was being financed on the basis of future oil production and the current economic situation would have been quite untenable without the existence of North Sea oil. This argument can perhaps be most easily understood in terms of the original Gregory figure (Figure 4.2).

The non-oil import demand function and the export supply function are taken to be stable. The economy at 1970 is taken to have been in equilibrium at $P_0 q_0$ where total exports equal total non-oil imports plus oil imports at 1970 oil prices. This equilibrium was disturbed in 1973 when the price of oil rose sharply. The new equilibrium would have then been at $P_1 q_1$ which would have given rise to a much higher level of exports, a lower level of non-oil imports and a much lower exchange rate. It would have required a considerable degree of structural adjustment to have reached this new equilibrium and as the government expected the output of North Sea oil to appear in the near future, it was decided to try and avoid this structural adjustment by borrowing from overseas to cover the temporary balance of payments deficit. So while some movement would have been made towards $P_1 q_1$, the economy had not in any sense adjusted to this new position. This is illustrated by the continuing balance of payments deficit between 1973 and 1978, despite the fact that oil production began in 1975 and was fairly substantial in 1976 (12.2 million tonnes). When

Figure 4.2: The Effect of North Sea Oil on the UK Economy Throughout 1970*

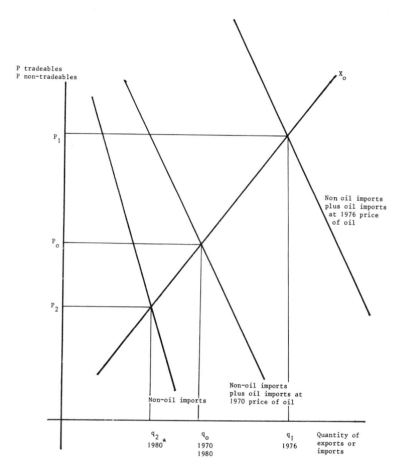

*The situation if North Sea profits, interest and debt repayments abroad are ignored.

the economy became self sufficient in oil in 1980 the equilibrium would have shifted again. If there were no foreign payments associated with oil production then the new equilibrium in 1980 would have been at P_2q_2 yielding a higher exchange rate than the original 1970 one. However, the Governor of the Bank of England points out that the total payments abroad caused by the North Sea sector (repayment of debt, profits going abroad, etc) are virtually identical to the cost of oil in 1970 so the actual 1980 equilibrium will be represented by a point almost identical to $P_0 q_0$, the 1970 starting point. There is, therefore, no need to make any major structural adjustment in the economy at all and there is no reason to expect the industrial base to decline either relatively or absolutely with respect to a 1970 base. The Bank took this argument further in an article in the March 1982 Bank of England Quarterly *Bulletin*. They argue that even if the North Sea had been developed solely from British investment, so that there would now be no outflow of capital, there would still be no net change in the trading position relative to 1970. The reason for this is that as the cost of extracting the oil at 1970 prices would have equalled or exceeded the value of the oil there would have been a need for a switch of resources from general production to the North Sea sector of a greater value than the existing oil imports. If, for simplicity, these resources came from the sector competing with general non-oil imports, there would have been a fall in production in this sector and a corresponding increase in non-oil imports so as to maintain consumption levels. In terms of Figure 4.2, the starting point would be $P_0 q_0$, North Sea production would reduce the total imports curve to the non-oil imports equilibrium $P_2 q_2$, but there would be a rise in non-oil imports caused by the fall in domestically produced substitutes so that the total non-oil import curve would move to bring about an equilibrium close to $P_0 q_0$.

Both the Bank and the Kay view are theoretically consistent. Deciding between them is essentially a matter of deciding how far the economy adjusted to the high oil price equilibrium $P_1 q_1$. If the adjustment was complete in 1976 then Kay's analysis is correct and a relative decline in the industrial base is inevitable. If the adjustment was in fact negligible then the Bank would have the more valid point of view and there would be no real need for any industrial decline. This may be most easily seen with the aid of a simple diagram. Figure 4.3 shows the time paths for the relative share of manufacturing in total output which the two views would give. Time path B represents the Forsyth and Kay view; manufacturing rises rapidly between 1973/4 and 1976 due to the effect of the oil price increase in 1973/4, then as domestic oil production begins this rise is reversed and the share of manufacturing in total output falls. Forsyth and Kay only analyse the fall which occurs between 1976 and 1980, of course, but their assumption of an initial equilibrium state implies that the 1973/4 to 1976 adjustment was completed by 1976. Path A represents an extremely simplified

Figure 4.3: The Contrasting Relative Share of Manufacturing in Output Given by the Forsyth and Kay View and the Bank of England

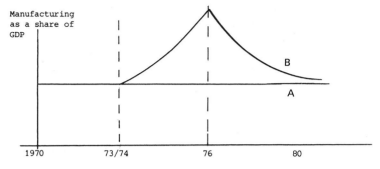

form of the Bank of England view. The manufacturing sector adjusts very slowly and so there is no discernible rise between 1973/4 and 1976. The economy is then out of equilibrium when domestic oil production restores economic equilibrium by removing the need for any structural adjustment. The key to whether any significant fall in the share of manufacturing needs to occur lies in the period 1973/4 to 1976.

If the economy had adjusted to the oil price shock by 1976 a number of things should have occurred. There would have been an increase in the volume of exports between 1974 and 1976, the visible trade balance should have returned to near equilibrium by 1976 and the exchange rate should not have continued its decline after 1976. In fact the volume of exports actually fell in 1975 and while the index for 1976 is higher than that for 1974 it is well below the level which would have been reached if the pre-1974 growth rate had continued (see Table 4.1). The visible trade balance for 1976 was −£3,927 million; this can hardly be described as an equilibrium situation. The inflation adjusted exchange rate continued to fall until 1978. The trade position certainly indicates that the industrial sector had failed to make a large part of the necessary adjustment by 1976. The general world recession may, however, have distorted the export situation, so it is also useful to examine the industrial sector directly. Table 4.2 gives a breakdown of industrial production through the 1970s.

For 1976 to be a valid base year for Kay's argument, industry in general and the manufacturing sector in particular should have reacted to the oil price rise by expanding its output relative to total GDP if not in absolute terms. In fact between 1974 and 1976 there was an approximate 5 per cent fall in both all industry production and

Table 4.1: The Trade Position of the UK, 1970-1979

	Volume of exports	Volume of imports	Visible trade balance (£m)	Invisible trade balance	Real effective[a] exchange rate
	1975 = 100				
1970	81.1	81.8	−32	+813	128.1
1971	85.9	85.5	+190	+886	132.1
1972	85.6	95.2	−761	+937	128.7
1973	97.2	108.4	−2,586	+1,530	116.5
1974	104.2	109.5	−5,350	+1,971	113.8
1975	100.0	100.0	−3,333	+1,659	114.9
1976	109.9	105.9	−3,927	+2,811	103.1
1977	118.4	107.4	−2,279	+1,995	101.7
1978	121.5	112.6	−1,546	+2,166	99.8
1979	125.9	125.7	−3,404	+1,541	108.3

Note: a. i.e. adjusted for movements in domestic and world prices.
Sources: Economic Trends 1981, Annual Supplement; authors' calculations.

Table 4.2: Index of Industrial Production (1975 = 100)

	All industries	All industries except oil and gas	Manufacturing
1970	99.7	99.7	98.0
1971	99.8	99.7	97.5
1972	102.0	101.8	100.0
1973	109.5	109.3	108.4
1974	105.2	105.2	106.6
1975	100.0	100.0	100.0
1976	102.0	100.7	101.4
1977	106.0	102.0	103.0
1978	109.8	104.1	103.8
1979	112.6	104.3	104.1

manufacturing output. It seems reasonable to conclude, therefore, that it takes the industrial sector considerably more than two years to react to a change in world market structures such as took place in 1973/4. The economy cannot, therefore, be thought of as being in equilibrium in 1976.

The debate between Kay and Richardson is essentially one about how actual past events should be interpreted. There is, however, another line of argument, which says that if the government had adopted expansionary fiscal policies throughout the middle and late

1970s the presence of North Sea oil would have allowed the economy to grow at a rate which would have prevented any possibility of an absolute decline in the manufacturing sector. This argument can also be explained within the framework of the Gregory model by making the import demand function dependent on the level of economic activity as well as the relative price of tradeable and non-tradeable goods. This change is an important one as it points to an implicit weakness in the Gregory-Kay argument. This is that overall income is assumed constant and the formal Gregory model seems to imply that income is being held constant at the full employment level. In fact the import function must shift with different levels of national income so that a higher income will generate uniformly higher import levels than a lower level of income. So the effect of a higher level of national income, in terms of the Gregory diagram, is an overall shift in the import function to the right.

The essence of this argument is that initially the economy is in a state of virtual balance of payments equilibrium ($P_0 q_0$) but that it is not at full employment. The level of national income consistent with balance of payments equilibrium is Y_1 given that the exchange rate is being held at the level implied by P_0. If fiscal policy were to be used to expand the level of national income to full employment, Y_f, this would shift the import function to $M_2(Y_f)$ and generate a balance of payments deficit as long as P_0 is maintained. This situation is believed to characterise the UK economy fairly closely during the 1960s and early 1970s. Whenever the economy began to grow and approach full employment the balance of payments would go into deficit and because the exchange rate was being maintained at a fixed level this would force the government to deflate the economy to cure the balance of payments deficit. The balance of payments was, therefore, acting as a constraint on the economy. It is suggested, therefore, that when North Sea oil output became significant the import function would fall to the non-oil import function (Y_1), as Kay has suggested. Kay's analysis essentially stops here with a rise in the exchange rate and a fall in the manufacturing sector. This view suggests that if fiscal policy were to be used to stimulate the economy the balance of payments would no longer be a constraint on growth and a higher level of income Y_2 could be sustained. In Figure 4.4, Y_2 is the level of income which maintains a balance of payments with P_0 held constant. The export sector will be unaffected and imports will be the same total quantity although their composition will have changed; there will be no oil imports but more non-oil imports. There will still be something of a relative decline in the

Figure 4.4: The Effect of Different Levels of Income on the Gregory Model

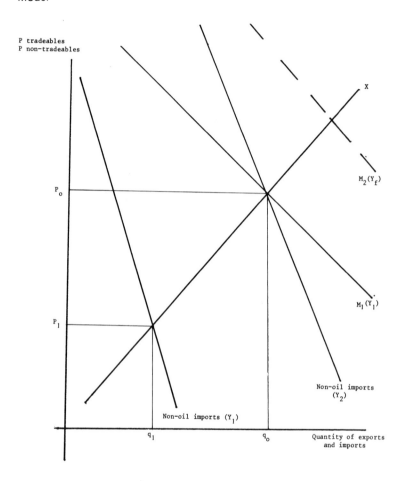

tradeable sector because total output will have risen from Y_1 to Y_2 while exports remain constant and imports rise. But there will certainly be no need for an absolute decline.

The flaw in this argument is basically the same as in Kay's own argument, that is, that the UK economy did not start to introduce oil production into an equilibrium situation. As Figure 4.2 shows, the effect of the 1973 oil price rise combined with the attainment of significant oil production levels operated to offset each other almost exactly. There is therefore no room to expand national income significantly without generating a considerable balance of payments deficit. Of course it is always possible to expand output and allow the exchange rate to fall to achieve a balance of payments equilibrium. But North Sea oil is quite irrelevant to this course of action as it was perfectly possible to do this in the 1960s before oil was even thought of. The reason why this course might be considered undesirable is basically because of its effect on inflation. There is no doubt that fiscal policy could be used to bring about near full employment and if the exchange rate were allowed to fall sufficiently this need not create a balance of payments problem. However, inflation would then be driven up through two distinct channels. Firstly, the high level of income would raise demand and encourage inflation; secondly, a low exchange rate would push up import prices which would inevitably feed through into inflationary effects.

These last points suggest reasons why this active fiscal policy approach may not have been acceptable in the UK during the 1970s. Nonetheless this argument for using fiscal policy to offset the effects of a natural resource on an economy is a very important one in the general context. It has been argued here that the basic Gregory analysis is correct in implying a relative decline in the tradeable goods sector when a major new natural resource is developed. If the economy is growing slowly then the relative decline may well be an absolute one as well. But if the government adopts an appropriate set of economic policies along the lines just discussed then the absolute decline in manufacturing can be prevented. In effect this says that a rather inappropriate assumption of the Gregory analysis and the simple Forsyth and Kay example is that government policy remains unchanged. This would be a totally inappropriate and unacceptable way for any government to react. Even if the arguments of the Bank of England did not apply to the UK during the 1970s, therefore, this argument would indicate that as long as government economic policies were correctly revised to allow for the advent of the oil there could be no absolute decline in manufacturing. If such a decline were to occur then it must be due to a failure in the government's economic policies rather than directly due to the oil.

One of the reasons why the Kay effect is so intuitively appealing is that it offers a fairly simple but consistent explanation for a phenomenon which is clearly affecting the UK economy. This is, that there is a decline in manufacturing industry. The promise associated with North Sea oil was that it would bring about a sustained real boom in the British economy. Instead of this promised growth in industry, output and consumption, the economy is clearly suffering a major recession with high unemployment, low levels of domestic investment and little prospect of a quick recovery.

It is very tempting, therefore, to blame North Sea oil for failing to live up to our hopes and therefore being the direct cause of the decline in manufacturing industry. This is not, however, a reasonable approach. Frank Blackaby (in summing up a collection of conference papers on *De-industrialisation* (F.T. Blackaby, NIESR, 1980)) has pointed out that 'in "de-industrialisation" we had a new label for an old problem — the relatively poor competitive performance of British manufacturing industry'. This gets very much to the heart of the problem, the decline in manufacturing in the UK is a very long-term one and as a long-term trend it clearly has nothing to do with the recent exploitation of North Sea oil. There is not scope to go into the arguments over the cause of the long-term decline in the UK industrial base here, the interested reader might profitably consult Blackaby's book, but Figure 4.5 makes the long-term trend in manufacturing quite clear. This figure shows how the percentage of manufacturing as a proportion of gross domestic product has changed between 1960 and 1980. There has clearly been a long-term decline from 36 per cent in 1960 to under 28 per cent in 1979. Between 1973 and 1976, the period in which Kay would argue that manufacturing expanded to adjust to the low exchange rate, high oil price situation, the proportion of manufacturing in total GDP actually fell. In 1977 and 1978, when the Kay effect would suggest that manufacturing should be falling as a proportion of GDP, it actually rose. Even a brief visual inspection of this figure makes it quite clear that there has been no major change in the pattern of relative decline as a result of the introduction of the North Sea sector.

The rise in the exchange rate in 1979/80, coupled with the decline in actual output levels (both GDP and manufacturing production) in 1980/81, also seems to fit very nicely into the Kay framework. This is, however, a grossly misleading suggestion; the relationship between oil and general economic policy will be discussed in the last three chapters where it will be argued that the major cause of this decline in production was the economic policy being pursued by the government. The highly deflationary fiscal policies which were adopted led directly to the fall in output. Insofar as there may have been a Kay effect operating at the same time, the government could have done a great deal to offset this effect by increasing demand levels. In fact the opposite approach was adopted and this must take the blame for the resulting developments, not the North Sea.

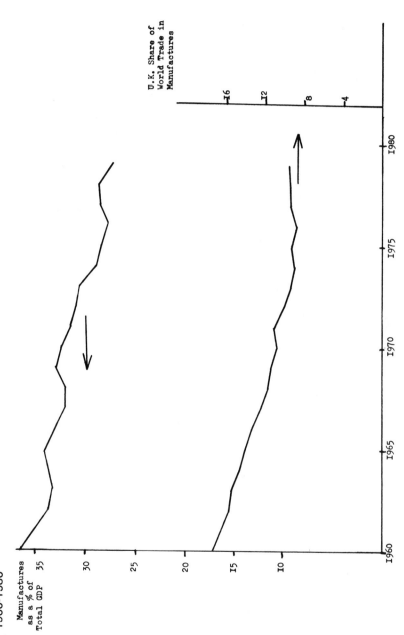

Figure 4.5: The Share of Manufacturing Output in Total UK GDP and the UK Share Of World Manufactured Exports, 1960-1980

To conclude, it is quite correct to state that the exploitation of a natural resource will cause an absolute decline in the domestic production of tradeable goods, if the economy starts from a position of balance of payments equilibrium and if nothing else changes. In the case of the UK economy these two qualifications do not apply; the large oil price rise in 1973-4 put the balance of payments situation well out of balance and real GDP rose by 20 per cent between 1970 and 1979. These two factors combined to make the Kay analysis far too simple to have any practical relevance to the introduction of North Sea oil into the UK economy and its effect on manufacturing industry. This, of course, does not mean that the oil has had no effect. But the Bank of England arguments make it clear that these effects have largely been to protect the UK economy from having to make large structural adjustments to the 1973/4 and 1979/80 oil price rises. It is not that we are much better off now than we were in 1970, but that we are much better off than we should have been without the oil.

5 THE EXPERIENCE OF OTHER COUNTRIES

In the last chapter it was argued that the Gregory analysis is correct as a piece of general theoretical analysis, even if not very relevant to recent events in the UK. It should be possible to interpret events in other countries in the light of this analysis. Within this context the two most important countries to be considered are Norway and the Netherlands. Both are Western industrialised countries which have shared in the hydrocarbon discoveries on the European continental shelf. They have adopted different lines of government action. Norway has aimed to maintain full employment through high levels of domestic investment, while the Netherlands adopted a broadly passive policy which allowed structural adjustments to take place. Australia has also been chosen for examination, partly because it is the home of the Gregory thesis but mainly because its own mineral boom came in the late 1960s, nearly a decade before North Sea oil began production. This means that the Gregory effect should have been in operation for much longer and so the long-term trend may be discernible. These three countries encompass most of the experience which is directly relevant to the UK; two other countries will also be considered, however: Japan and Venezuela. Venezuela has been chosen because it is an example of a country which, for a very long period, has been a major oil producer. Japan is of interest because it is an economy largely lacking in oil and other energy resources and so it is likely that a reverse Gregory effect will operate. That is to say, a large world price increase of a natural resource will affect an economy such as that of Japan in exactly the same way as a loss of natural resources. In terms of the simple Gregory analysis outlined in the last chapter, the effect of an increase in world oil prices on an oil importing country would be to reduce the exchange rate and stimulate the production of tradeable goods.

The objective of this chapter is not to provide an up-to-date survey of these five countries. Such surveys are available from a number of sources such as the OECD and the National Institute Economic Review, which are not subject to the same order of publication delays as must affect a book such as this. The objective is simply to chart the development of the five countries over a fixed period during which three of them were affected by natural resource discoveries and all five were affected by the large increase in oil prices in 1973/4. The period which has been chosen is 1970-80.

Norway

Norway is often presented as an ideal example of the correct exploitation of natural resources in a manner which has completely protected the non-oil economy. The 1981 TUC economic review cites Norway as an 'example of the constructive approach to North Sea revenues which the TUC wants Britain to adopt'. It is true that a quick appraisal of the Norwegian situation seems to show relatively little industrial decline and none of the exchange rate effects which Gregory predicts. On this basis many people have called for the adoption of policies similar to those of Norway in the belief that similar benefits will result in the UK. The theoretical reply to this would be that output and employment in the tradeable sector can be maintained by investing heavily in it, but this may be done ultimately at the expense of domestic inflation and a loss in the consumption which might have resulted from the oil sector. What is perhaps worse, the investment in the tradeable sector will tend to be in support of the weaker industries and will probably not show a realistic return.

Norwegian exploration in the North Sea began in 1966, and the first discoveries were made in the late 1960s. Oil production began in 1971, four years earlier than in the UK. By 1980 oil and gas production represented 15 per cent of the Gross Domestic Product. Output remained quite low between 1971 and 1974, it then increased rapidly after the completion of the Emden and St Fergus gas pipelines in 1977. The onset of production from the Statfjord field in 1979 brought about a rapid rise in oil output and total production was close to 50 million tonnes oil equivalent in 1980. While this figure is well below the comparable UK one, some idea of the overwhelming impact of oil on the Norwegian economy can be gained from the fact that total domestic consumption of oil products is less than 9 million tonnes oil equivalent per year. Thus Norway is the largest net energy exporting country of the OECD with an annual oil and gas surplus in excess of 40 million tonnes oil equivalent. The oil and gas sector is currently providing about a third of total government tax receipts. In the longer term the Norwegian Ministry of Petroleum and Energy sees a steady rise in production through the rest of this century, reaching approximately 90 million tonnes oil equivalent in the year 2000.

The fact that the onset of large-scale oil production in Norway virtually coincided with the increase in oil prices in 1973/4 largely protected the Norwegian economy from making any adjustment to the oil price rise as such. Unlike the UK, the Norwegian exchange rate did not fall despite the increase in the balance of payments deficit which occurred from 1973 onwards. In fact, from 1970 to 1977 there was a steady, but quite definite, upward trend in the value of the krone. Total industrial production and the Gross Domestic Product have both shown a healthy growth rate over the whole period. The unemployment rate peaked in 1975 at under 2.5 per cent and was well under 2 per cent for the rest of the decade. There has been little sign of a rise in the overall inflation rate over the decade. The rate of change of consumer prices peaked at under 13 per cent in 1975, then steadily declined until late 1979 when it stood at only 4 per cent. It then rose rapidly to peak at 14 per cent in late 1980 and early 1981. The current account of the balance of payments was in deficit for the whole period except for a small surplus in 1972 and a large surplus in 1980 due to the large increase in oil production. The surplus on the current account is likely to persist during much of the 1980s, due to the large contributions from the oil and gas sectors.

A brief glance at the Norwegian situation seems to show an economy with considerably less unemployment than the other industrialised countries in spite of a slowly rising exchange rate and a newly developed oil sector. It certainly seems that oil has had only beneficial effects within the economy.

The explanation lies in three distinct areas; first the majority of Norwegian exports are founded on a considerable natural advantage, second the direct effect of oil, and third, direct government intervention. In 1980 Norway's largest export, by total value, was oil itself, the next largest was aluminium, then shipping, fish and paper. In all these areas Norway has considerable natural advantages which make it able to withstand a fairly substantial decline in its competitive position. Aluminium production, for example, involves huge quantities of electricity, which Norway can supply very cheaply because of its large hydroelectric resources. In terms of the Gregory diagram it is as if the export supply function were nearly vertical; the new oil sector changes the relative prices by raising the exchange rate, but the quantity of goods traded alters only slightly.

This strength in the traditional exporting sectors of the Norwegian economy does not, however, explain very much of the growth in GDP which has occurred. This is explained by the second important factor, which is the enormous direct contribution which oil and gas have made to the Norwegian national accounts. Almost all the growth which has occurred both in GDP and in industrial production has come from the North Sea sector itself. The vast majority of the traditional sectors have shown almost no growth since the early 1970s.

The last major factor which has contributed to the apparent success of the Norwegian economy is the government's own policies. Under the influence of a slowly rising exchange rate and the growth of such a major new exporting sector, we would normally expect to see a fairly substantial reorganisation of the structure of the domestic economy. This would normally give rise to at least a transitory rise in unemployment. This has not occurred in Norway, primarily because the main objective of government policy has been to prevent any rise in unemployment. This has been achieved largely by making direct grants and subsidies to the industrial sectors so that less profitable processes may be continued. Table 5.1 shows the very large increase which has occurred in the level of government grants and subsidies. This is contrasted with the rise in tax revenues from the oil sector, and it can be seen that a very large part of the tax revenues, even in 1980, is flowing into maintaining less profitable industries through direct grants and subsidies. Figure 5.1 shows the export performance of Norwegian manufacturing industries; even with government aid the manufacturing sector has rapidly been losing its share of world markets.

Table 5.1: Norwegian Government Subsidies and North Sea Tax Revenues (Kr billions 1975 prices)

	Grants and subsidies	Total North Sea taxes
1972	6.8	0.06
1973	7.6	0.1
1974	8.2	0.2
1975	9.3	1.4
1976	10.9	3.0
1977	12.2	2.9
1978	13.3	4.9
1979	12.7	8.0
1980	12.8	15.0

Source: OECD.

This attempt to maintain the level of industrial activity through direct investment and support by the government has only been partly

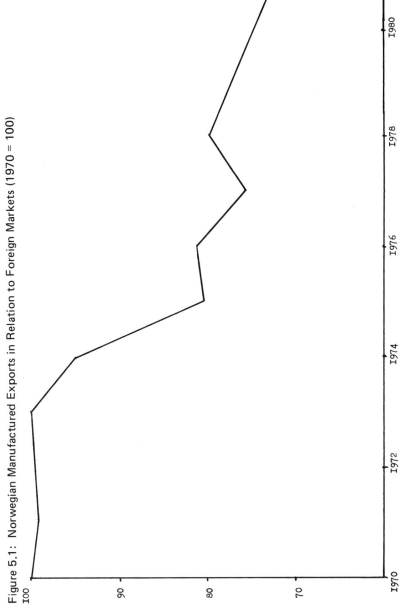

Figure 5.1: Norwegian Manufactured Exports in Relation to Foreign Markets (1970 = 100)

successful. Employment has been maintained at a high level, but export performance has been poor. The important question, however, is what has happened to Norwegian industry generally. A decline in exports need not be a bad thing if growth levels have been maintained and products have simply shifted to domestic consumers. This is not, however, the case. Despite the level of government support the manufacturing sector has shown an absolute decline since the early 1970s. In 1974 the GDP contribution of manufacturing was Kr33,280 million (1975 prices), in 1980 its contribution was Kr32,513 million (1975 prices), a fall of nearly 2½ per cent. Similarly the contribution of agriculture, forestries and fisheries was Kr9494 million (1975 prices) in 1974, and this had fallen to Kr8,938 million (1975 prices) by 1980. The only sectors which had shown a large rise were those related to the North Sea itself and some of the service sectors, particularly retailing and social services.

If non-oil output remains almost constant during a time of high employment this must mean that there is virtually no increase in productivity. This implies that the large quantities of investment funds flowing into Norwegian industry are not being used to build a more effective and competitive industrial base. Rather, the funds are simply supporting a very high degree of hidden unemployment and protecting industry from having to make some painful but necessary structural adjustments. This weakness of productivity and general industrial efficiency is illustrated in Figure 5.2.

This situation of a steadily rising exchange rate causing poor export performance coupled with high levels of employment and demand causing rising imports is one which cannot continue indefinitely. This was recognised by the Norwegian government towards the end of 1977 when economic policy underwent a fairly radical reorientation aimed mainly at restoring international competitiveness and external balance. This took the form of a generally more restrictive monetary policy, a prices and incomes freeze from the end of 1978 to the end of 1979, and a devaluation of the krone in February 1978. The general deflation of the economy has been very mild, however, and the aim of maintaining full employment was not abandoned. There has been an attempt to reduce the support to industry from 1977 onwards, although the 1980 figure is still well above the 1973 figure. The non-oil export performance has continued to decline, if a little more slowly, but the balance of payments did move into credit in 1979 and 1980, due to increases in oil production and exports. It seems clear that a policy of direct investment to maintain the industrial base of the country has not

Figure 5.2: Labour Productivity in Manufacturing Industry

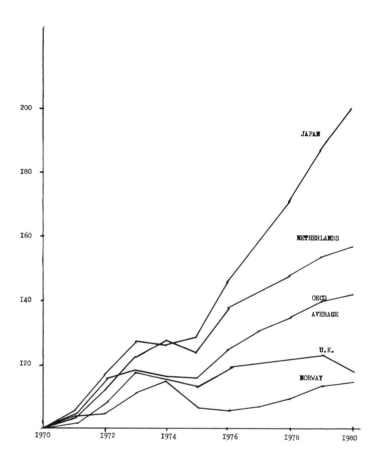

Source: OECD.

succeeded in building a more competitive and viable industry in the Norwegian case. Unemployment figures have remained at a low level, but this seems to be at the expense of a poor trend in productivity.

Given that much of the oil resources are channelled towards supporting industry and maintaining employment level it seems reasonable to expect that consumption will be well below its potential level. It is hard to establish if this has occurred. However, if we compare the Norwegian and Netherlands cases, we see that in the early 1970s

Norway enjoyed a slightly higher level of consumption than the Netherlands, but by the late 1970s the situation has been reversed.

Supporters of the Norwegian case would argue that this loss in consumption is merely the price which is being paid in order to maintain a healthy industrial sector. They argue that this industrial sector will be viable when the oil is exhausted and Norway is once again solely reliant on its non-oil sector. The question which this raises is how will Norwegian industry be able to survive without its oil subsidies when the oil has gone (although this may well be 50-100 years into the future). If the industry is really maintained in a viable state only awaiting the lower exchange rate of the non-oil economy, then of course it will be able to survive. If, on the other hand, industry becomes increasingly inefficient and labour-intensive, because of the attempt to maintain employment and prevent structural change, then the end of the oil may also herald the decline of the non-oil sector.

The Netherlands

In contrast to the Norwegian case the Netherlands is often presented as a prime example of the mismanagement of a national resource. The picture which is often given of the 'Dutch Disease' is one of an economy which has embarked on a massive consumption boom, using the wealth of its natural gas sector to finance large quantities of imported consumption goods. The domestic non-gas industrial sector is seen to be in sharp decline, due partly to insufficient investment and partly to the effects of a high exchange rate reducing the price of imported consumer goods.

At a superficial level it is quite easy to find support for this view in the statistics for the Netherlands. The exchange rate has risen steadily over the period 1970-80 from an effective rate of 88.0 (IMF index) to 119.6, an increase of almost 36 per cent. This has accompanied a huge increase in imports; in constant prices the value of total imports rose between 1970 and 1980 by 55 per cent. The unemployment rate is usually presented as an indication of the industrial stagnation which has occurred. In 1970 unemployment stood at 1.2 per cent; this figure rose steadily, with only a slight break in 1977, to 5.3 per cent by 1980.

This rise in unemployment is a considerable one in a country which has traditionally enjoyed low unemployment figures. The vast growth in the social security sector and in general consumption levels also supports the idea that the benefits of natural gas are being wasted on a once and for all spending spree. The concept of the 'Dutch Disease' and the dire warnings which often go with it are a clear result of this view.

But how fair is the overall view of the Netherlands which this approach implies? The Netherlands government's main policy approach during the 1970s, and particularly the time of the first oil price rise, was largely a macroeconomic one in the sense that they have used fiscal policies to maintain overall demand levels but they have not intervened to a large extent directly in industry. This has led to a high general level of consumption and demand and a steady rise in the exchange rate has occurred. The rise in the exchange rate has accompanied a considerable shift in the trading patterns of the Netherlands. Natural gas has become a major part of the export sector and traditional exports have declined in importance; similarly a large degree of import penetration has occurred. This picture of a relative decline in non-gas exports, an increase in imports and an increase in unemployment leads many people to conclude that manufacturing industry as a whole must be suffering a decline.

This is in fact not the case; between 1970 and 1980 manufacturing production in the Netherlands has increased by 28 per cent compared with an increase of 17 per cent in Norway, 24 per cent in Germany and virtually no increase in the UK. There have been major structural shifts within manufacturing industry, and those areas which are least able to compete on world markets have contracted considerably. This is particularly the case in the textile and clothing industries.

But other industries, such as chemicals, iron and steel, electrical appliance manufacture, have been growing rapidly. This is illustrated in Table 5.2 which shows an all industry increase of 38 per cent between 1970 and 1979. Most of the individual sectors have in fact suffered a relative decline — manufacturing, food and beverages, printing, etc. — but the overall growth in the economy has allowed this relative decline to take place in the context of a considerable absolute growth. Only a very few of the tradeable sectors have shown an absolute decline. Leather goods and textiles are by far the most dramatic case. It is, of course, to be expected that this sector would be particularly vulnerable as many of the Third World countries are making large inroads into these trading sectors.

At the same time as the general increase in unemployment there has been a fairly considerable change in the nature and structure of employment in the Netherlands. There has been a very large movement of labour away from the production sectors of the economy and into the

Table 5.2: Index Numbers of Industrial Production for the Netherlands (1970 = 100)

	1970	1975	1977	1978	1979	1980
Industry total	100	123	133	134	138	137
Mining & quarrying	100	238	245	226	238	221
Manufacturing	100	109	119	121	125	128
Food, beverages, etc.	100	119	127	132	135	136
Textiles	100	81	81	77	79	74
Leather goods	100	62	61	57	57	56
Printing & publishing	100	104	115	120	123	125
Chemicals	100	125	151	156	170	164
Basic metal industries	100	100	107	114	115	112
Electricity, gas, water	100	153	164	175	178	175

Table 5.3: The Structure of the Labour Force and Employment in the Netherlands

	1970	1971	1972	1973	1974	1975	1976	1977	1978	1979	1980
Total employment 100	4585	4612	4569	4576	4578	4552	4547	4551	4577	4616	4605
Unemployment[1] %	1.2	1.4	2.4	2.4	3.0	4.4	4.6	4.5	4.6	4.6	5.3
% Employment in agriculture mining and quarrying and public utilities	3.7	3.5	3.4	3.3	3.1	3.1	3.1	3.1	3.1	3.0	3.0
% Employment in manufacturing	30.0	29.3	28.6	28.0	27.8	26.9	25.8	24.9	24.0	23.3	22.9
% Employment in construction	11.5	11.2	10.8	10.7	10.2	9.9	10.0	10.0	10.2	10.3	9.9
% Employment in commercial services	34.0	34.4	34.6	34.7	34.8	35.1	35.3	35.5	36.0	36.3	36.5
% Employment in government and other services	20.7	21.6	22.6	23.3	24.0	25.0	25.9	26.5	26.8	27.0	27.6

Source: OECD Economic Survey of the Netherlands 1981
[1] Unemployment includes individuals on public work projects

service sector. Table 5.3 shows the large fall which has taken place in manufacturing employment. This fall is almost completely offset by the rise in government and non-commercial service employment.

This combination of falling employment levels in industry combined with general increases in industrial output gives a fairly considerable increase in labour productivity. Figure 5.2 shows the trend in productivity during the decade of the 1970s.

Considerable structural change has occurred in manufacturing industry, but because of high levels of domestic demand this has not led to an industrial decline. Instead, large increases in productivity have occurred and industry generally is more able to compete effectively on world markets. What, then, of the charge that all the gas revenues are flowing into consumption? This contains an element of truth; consumption, both private and government, has increased considerably. The share of government expenditure has also risen sharply. In 1970 general government expenditure represented 52 per cent of net national income, by 1981 this figure had reached 67 per cent. By far the largest factor in the increase has been the growth in the general social security services. But this is also by no means the full story. Up to 1971 the Netherlands had traditionally been a net importer of long-term capital, but from 1971 onwards this changed, especially after the price of oil rose in 1973, and the Netherlands undertook considerable net foreign investment. Indeed by 1978 the Netherlands had become the largest single direct foreign investor in the United States of America. This flow of investment abroad has been positively encouraged by the government which removed barriers to capital movements and maintained a policy of low interest rates. This foreign investment was mirrored by the large balance of payments surpluses from 1972 until 1979.

The popular conception of the 'Dutch Disease' proves to be wrong on two counts. Firstly, the Netherlands has invested a large amount of the gas revenues and secondly, there has not been an industrial decline as such, merely a redistribution in production and an increase in productivity resulting in a rise in unemployment. To a large extent the changes which have occurred in the Netherlands industrial sector are desirable. The older, less efficient, industries have declined while newer ones have been given a chance to grow and develop in an environment of plentiful demand and investment capital. The only disadvantage of this structural change is the large quantity of unemployment which has resulted. Some of this unemployment will disappear into all the new industries, but inevitably some people will be either too old to be retrained or, because of the higher productivity of the new industries, simply not needed. The Netherlands have responded to these cases by constructing a social service system which is arguably the best in the world and is capable of maintaining standards of living even if not of providing jobs.

Gas is likely to play a major part in the Netherlands economy well into the 1990s and so talk of re-entry problems is obviously premature. On present trends, however, it seems likely that the Netherlands industrial sector will be very well placed to take advantage of the ultimate fall in the exchange rate when North Sea gas disappears.

A Comparison of the Policies of Norway and the Netherlands

Norway and the Netherlands provide a particularly interesting comparison from the UK viewpoint; both countries are highly industrialised and experienced the exploitation of major hydrocarbon discoveries at about the same time. In many ways the Netherlands is a closer parallel to the UK than Norway. Its rate of gas exploitation has produced a sector which is closer in relative size to the UK case than the Norwegian. The industrial base of the Netherlands is also more like that of the UK in the sense that it lacks any of the major Norwegian advantages such as hydroelectricity. Nonetheless, both countries are sufficiently similar to the UK to justify a fairly close comparison. This can be particularly instructive in view of the markedly different policies which the two governments have pursued. The Norwegian aim has been to maintain full employment through direct aid to industry, while the Netherlands have pursued a policy of maintaining high levels of demand and a generous social security system.

Neither policy has prevented the relative decline of traditional tradeable goods sectors, but their effects in other ways have been quite markedly different. The Norwegians have certainly been successful in their aim of maintaining high employment. But this success has been at the cost of an almost total stagnation of their non-oil industrial sectors. The sharp reorganisation of the domestic economy which one would expect, and the decline of some of the traditionally less competitive industries while others respond positively to the increased domestic consumption levels, has simply not occurred. By using much of the oil wealth to prevent the decline in these industries which are not really viable, the government has discouraged the high demand levels which might have stimulated other sectors into rapid growth.

In the case of the Netherlands the expected economic reorganisation has been directly encouraged by the government's policies of fiscal expansion without direct industrial support. This has produced growth rates within many sectors of the economy which are remarkably high. Between 1970 and 1979 production in the chemical industry increased by 70 per cent, and electricity, gas and water increased by 78 per cent, although some sectors have shown a large decline. The cost of this reorganisation has been felt most strongly through the rise in the unemployment rate, although by some standards unemployment of the order of 6 per cent is not very high.

It is not possible to decide which of these alternative policies has been most successful without first making some largely subjective value judgements. If the maintenance of high employment is the primary concern of the government then the Norwegian policy has been clearly more successful. If, however, a greater weight is placed on industrial performance and the construction of a sound industrial base, then the Netherlands would seem to have adopted the better policy.

It is also necessary to consider the longer-term consequences of these two policies. Oil production is likely to continue in Norway for a very considerable period. This means that there is no foreseeable time when the situation will return to the pre-oil one. The industries which are currently absorbing such large grants will therefore have to be supported for a very long period. Much of the potential benefit of the oil would seem to have been used up to support these industries, and until they are allowed to decline it is hard to see this situation changing. Industrial support might be a sensible policy, therefore, if the new resource is likely to last only a short period of time, but if the natural resource is likely to last a very long time then direct industrial support would seem merely to postpone the necessary changes and costs of change.

Australia

Gregory's original paper, outlining the effects of natural resource exploitation on the rest of the economy, was written in consideration of two particular factors in the development of the Australian foreign trade sector. The first of these was a major boom in the Australian mining industry in the late 1960s. The second was the widespread use of tariffs to foster a domestic manufacturing sector. The mining boom was based on three distinct sectors: iron ore, bauxite and coking coal. Australia is currently the world's largest producer of bauxite, and the second largest exporter of coal. In the middle sixties the Australian people were promised a rapid expansion of industry, consumption and welfare levels based on the new mining sector. In December 1975 Mr Gough Whitlam, the Labour Prime Minister, was defeated by Mr Malcolm Fraser's coalition party largely because of the failure of the promised boom to materialise.

To a large extent the Australian case is a prime example of the Gregory effect in action, up to 1975. The value of the early seventies mineral boom was somewhat smaller than North Sea oil in relation to the total economy. It is obviously hard to value precisely a resource boom which is widely spread across a large geographical area and consists of a number of different resources, but the total GDP contribution of new resources was probably only of the order of 2-3 per cent of GDP. However, because external trade is smaller relative to the rest of the economy in the Australian case than in the UK, the effect on the balance of payments of the new resources was very similar to the UK, perhaps in the region of 25 per cent of the value of total exports. As the value of the resource began to mount through the late 1960s and early 1970s, the exchange rate rose steadily from $1.11 US per Australian $ in 1969 to $1.488 US per Australian $ in early 1974. Both exports and imports increased and there was a tendency towards a balance of payments surplus. The overall growth in the economy was sufficient to maintain employment levels, although relative shifts were occurring in the structure of the economy. The main tradeable sector in the Australian economy was traditionally its agricultural exports and this sector certainly showed a relative decline. The overall effect of the resource from 1970-4 seems to have been a relatively painless structural shift in output at a time of comparatively high employment and consumption levels.

In 1974/5, however, the pattern changed dramatically; the exchange rate began to fall, employment fell, the rise in consumption slowed considerably and the balance of payments moved into deficit. It seems, therefore, that the Gregory effect was only a temporary one associated with the introduction of the new resources. Gregory himself rejected this suggestion at a conference in London organised by the Institute of Fiscal Studies (1980) but he was unable to offer an alternative explanation for the sudden change.

In the light of the previous chapter, the explanation seems to be fairly clear. The sudden change in 1975 was exactly what Gregory's analysis would predict as the result of the increase in the price of oil imports. In effect the Australian situation is very similar to that of the UK. The new resource acted as a buffer protecting the economy from having to make a major adjustment to the new oil price situation. The only difference is that in the UK, North Sea oil did not actually begin production until after the oil price rise, so the positive effect of North Sea oil was never felt. In Australia the new resources began production much earlier and so what first appeared as a boom, in the early seventies, was cancelled out by the change in real oil prices and the slump in iron ore prices caused by the recession in the steel industry. This had the appearance of a sudden collapse in the boom itself; in fact the benefits of the resource were just as powerful after 1975 as before, but after 1975 they were acting to prevent a collapse in consumption levels rather than to maintain a positive increase.

It is interesting that the government in Australia is once again predicting a new mining boom through the 1980s. This time the boom is to be based on energy resources, steaming coal, natural gas and uranium. Once again it is predicted that there will be increased welfare in terms of consumption and employment and providing a new impetus to the whole economy. It will be interesting to see whether this new boom actually emerges as a positive effect on the Australian economy or if, once again, it is swallowed up by the huge increases in oil prices which occurred in 1979 and 1980.

Japan

The Japanese economy is singled out from most other economies by its almost total reliance on imported primary energy sources. The USA imports about 19 per cent of its basic energy needs, the UK in the late 70s imported about 35-45 per cent, West Germany about 55 per cent and France 80 per cent. In contrast to these traditionally energy-dependent states Japan imports close to 90 per cent of its primary energy needs. Approximately 7 per cent of its energy comes from nuclear power or hydroelectric power and there is a very small amount of indigenous hydrocarbon production. The Gregory analysis would suggest, therefore, that the Japanese manufacturing sector would be considerably stronger than that in most other manufacturing countries. It would also suggest that the Japanese economy would be particularly vulnerable to changes in the price of primary energy products.

The general strength of the Japanese trading sector is so well known that there is no need to dwell on it. Since World War II the Japanese have undergone a massive industrial revolution which has transformed their manufacturing sector; their industry has made major inroads into many of the world's markets in manufactured goods. This rapid growth is not, of course, solely due to the Gregory effect. In the early postwar years, low labour costs gave Japan a major advantage. Moreover, industrialising after World War II allowed Japanese industry to be sited and set up in a way which best suited modern techniques. However, Japanese industry has been operating in an international price environment which is very favourable. The yen has often been said to be undervalued, and the continuous balance of payments surpluses which Japan has enjoyed testify to this. In 1979 oil imports represented nearly 40 per cent of the value of total imports. These imports are therefore acting to reduce the effective exchange rate of the yen and maintain the price advantage of Japanese manufactured goods. There can be no doubt that if Japan were suddenly self-sufficient in oil, a major revaluation of the yen would occur and Japanese industry would largely cease to enjoy this major international advantage.

From 1970 to 1973 the Japanese economy was fairly stable, employment was high, and there was a constant, moderate balance of payments surplus which was causing a steady rise in the exchange rate. Investment was stable at a fairly high figure and *per capita* consumption was rising quite rapidly. The sudden increase in oil prices in 1973/4 showed itself initially in a large jump in the value of total imports. The Japanese government reacted very quickly to the change in import prices by adopting sharp deflationary measures. In 1972 the discount rate stood at 4.25 per cent; by the end of 1973 it had been raised to 9 per cent. Real GNP fell between 1973 and 1974, although a combination of domestic deflation and a substantial fall in the exchange rate caused real exports to rise by 18 per cent so that the balance of trade deficit which occurred in 1974 and 1975 was quite small. Partly because of the rise in oil prices, and partly because of the government's response to it, there was an actual fall in nominal consumption levels between 1974 and 1975. The reaction of the Japanese government to the oil crisis is interesting as it underlines a point which will become important in the last three chapters. This is the idea that a government's economic response to a sudden change in sector such as North Sea oil or world oil prices cannot be decided in isolation from general economic events and the view which the government takes on how the economy operates. The crucial factor in understanding the strong response which the Japanese government made to the 1973/4 oil crisis lies not in the oil sector but in the behaviour of the inflation rate immediately before the rise in oil prices. The rate of increase of the consumer price index throughout the whole of 1971 and the first half of 1972 was very close to zero. During the second half of 1972 and into 1973 the rate of increase of the index began to rise quite rapidly reaching 17 per cent in the third quarter of 1973. The increase in the oil price came on top of this already rapidly rising general inflation. It was this existing general inflation which made the Japanese government react so harshly to the oil crisis as it considered any worsening of the situation to be unacceptable. The rate of inflation peaked in the first half of 1974 at 35 per cent and then rapidly declined to under 5 per cent by the second quarter of 1975. A combination of the direct effect of the rise in oil prices and the sharp government response brought about a very rapid switch of production into the export sectors.

In 1975 the discount rate was reduced to 6.5 per cent, real national income and consumption began to grow at something like their pre-1973 rates, and from 1976 to 1978 the balance of payments moved heavily back into surplus. In effect it seems that the adjustments have occurred almost as a discrete change in the levels of variables without affecting their growth rates. So the exchange rate had been rising steadily at about 10 per cent a year; it suddenly fell in 1974, remained stable to 1976 and then began rising again. Consumption had been rising until 1974, it fell dramatically in 1975 (in real terms) as the Gregory thesis predicts, and then continued to rise. The Japanese government seems therefore to have acted in a way which brought about the structural changes even more quickly. It may well be that because of this long-term effects on the economy have been greatly reduced.

The oil price rises of 1979 and 1980 have once again led to a similar train of events. The large balance of payments surplus of 1978 has been turned into a deficit, the discount rate was again raised to 9 per cent during 1980, having fallen to 3.5 per cent in 1978, and the effective exchange rate has again fallen. It is too early to tell if the overall adjustment will be as swift and effective as in 1974/5 but the overall pattern is certainly remarkably similar.

Venezuela

Venezuela is the oldest oil producing member country of OPEC. Oil production started, on a small scale, in 1917. It is a founder member of OPEC and exerts a major influence over OPEC policies. The Venezuelan oil reserves were estimated at around 2,000 million tonnes during the early 1970s. At 1970 production levels (193 million tonnes) this would have been exhausted by 1981. The government therefore seemed to be faced by a major crisis as the non-oil economy is very weak and could certainly not survive in any recognisable form without the oil sector. Oil exploration had been traditionally carried on by the large oil companies in Venezuela. From the 1920s through to the late 1950s, however, the oil companies became aware of the increasing probability of nationalisation and ceased to carry out any further major exploration, but instead concentrated on extracting as much of the oil as possible. In fact nationalisation did not occur formally until 1 January 1976, when Petroleos de Venezuela, the national oil company, was founded. It faced a serious problem of dwindling proven oil stocks, which was only slightly relieved by a rapid reduction in depletion rates. By 1978 oil production had been reduced to only 108 million tonnes, almost half the 1970 figure.

New exploration is being undertaken in three distinct areas; off-shore exploration is being undertaken at a number of sites, particularly the Orinoco river delta. This can involve drilling far out to sea, in some cases at distances comparable with the North Sea rigs, although the climatic conditions are much milder. Early estimates of the offshore reserves are very promising (1,000 million tonnes to 7,000 million tonnes) and put the Venezuelan offshore sector on much the same scale as the North Sea. In addition to this, Venezuela is initiating a major programme of onshore geological surveys and drilling exploration. The fact that the success ratio on these new exploratory wells has been around 40 per cent shows that the onshore reserves are far from exhausted. The third area of expansion is the Orinoco heavy oil belt. This is an area of heavy oil deposits which are much more difficult to exploit as the oil is not easily extracted and needs a considerably more complex refining process.

However, the size of the reserve is so huge that as a long-term prospect it is immensely valuable. Estimates of the potential oil reserves range from 100,000 million tonnes upwards. This resource alone could maintain Venezuelan levels of production well into the next century. Thus there has been a complete transformation in the outlook for Venezuelan oil production since the early 1970s. Venezuela can now look forward to secure oil production levels well into the next century and there is really no need to consider an end to oil production at all.

The importance of oil to Venezuela is much greater than in the case of the European oil producing countries. Oil output is in the region of 30-40 per cent of total GDP, and it is therefore the primary source of domestic consumption. In the early seventies the economy was, broadly speaking, in a state of equilibrium; it was running a balance of payments surplus, consumption was growing at a moderate rate, although it was low by international standards, and domestic investment levels were fairly stable. Oil output was being slowly reduced because of the low levels of proven oil reserves. In 1973/4, when the price of oil rose dramatically, the initial impact was on the value of exports and GDP. Between 1972 and 1974 the value of exports rose from 17.4 billion Bolivares to 66.2 Bolivares, despite the fact that oil output actually fell. This huge increase in foreign earnings was initially far beyond the absorption capacity of the country, and so the balance of trade surplus produced a build-up of large quantities of foreign assets. The Venezuelans dealt with this in two ways; firstly, the huge increase in export revenues gave them the opportunity to cut their oil production levels considerably without affecting their ability to import. Secondly, a number of large investment programmes were set up on the basis of imported investment goods so that imports gradually rose from 1973 onwards. The percentage of domestic investment to GDP fell in 1974, not because actual investment fell but because GDP rose so suddenly. Actual investment then rose very quickly so that it reached a peak of 40 per cent of GDP by 1978. As imports and investment rose the increase in oil wealth also fed through into total consumption.

The main interest of Venezuela to this book does not, however, lie in its oil sector; it rather lies in the rest of the economy and the effect which this mammoth oil industry has had on it. The non-oil economy tends to exist on two distinct levels, a major industrial level which is heavily government subsidised, such as large steelworks, a major railway development, the construction of an underground railway in Caracas, etc., and a very low level of local services and production. In effect, the whole middle sector of the economy is missing: the sector which normally produces goods in competition with the import sector. There are virtually no small to middle-sized privately owned industrial units.

Even the agricultural sector seems to be unable to compete with imported goods. In 1971, Venezuela imported 46 per cent of its basic foodstuffs; in 1977, a fairly good year in agricultural terms, this figure had increased to 63.8 per cent. This trend is in spite of a deliberate policy on the part of the government to build up the agricultural sector through high levels of grants and investment in agriculture. The government and professional organisations are generally puzzled by the effect; some suggest that it is due to an inefficient economic infrastructure which is unable to provide the correct support for the farms. Others argue that it is due to the farmers themselves being unable to make efficient use of the new equipment, or general corruption preventing the funds from reaching those who need them. A more basic answer is perhaps that an extreme form of the Gregory effect is at work. When imports are cheap and imports are penetrating into the economy so effectively this means that the import prices are keeping all domestic prices down and reducing the incentive to produce domestically.

It is almost impossible to see how this situation can be changed, especially as the sector affected is as basic as agriculture. If, for example, the government wanted to build up a small electrical goods industry it could do so by banning imports of these goods and allowing a domestic industry to grow up. This would mean a severe shortage of these goods while the domestic sector grew and protection would have to be maintained until oil production declined, but this is a possible course of action. In the case of agriculture, however, this is not feasible as the shortage of food and the subsequent rise in food prices would cause massive hardship amongst Venezuela's poorer population. As things stand the Venezuelan government is trying to make the agricultural sector grow by investing heavily in it, despite the fact that there is no demand for these funds in the industry. Farmers are therefore often being persuaded to buy machinery and implements which they have little intention of using and virtually no incentive to maintain. These large amounts of investment would therefore be more profitably employed elsewhere, perhaps, following Kay, overseas where they would provide a source of income for agricultural support if and when the oil sector does disappear.

Venezuela can therefore be seen as an almost perfect example of the Gregory thesis taken to its extreme. As long as Venezuela can maintain its oil production levels, and at present this seems to be almost indefinitely, there is really no reason to try to build up domestic output in the face of overwhelming foreign competition. Investment should instead go either into the non-tradeable sector, services, social infrastructure, etc., or overseas.

Conclusions

The first, and most obvious, conclusion which emerges from this brief examination of five countries during the 1970s is the wide applicability of the Gregory analysis. The sudden oil price increase of 1973/4

Figure 5.3: Per Capita Consumption 1970-1980 in USA dollars

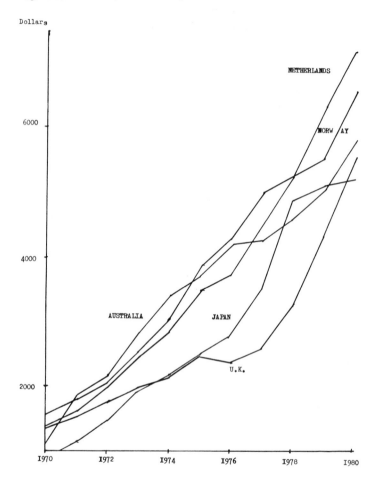

Source: OECD.

had repercussions on almost every state in the world, and these effects
can largely be analysed within this fairly simple theoretical model. The
general predictions of the model seem to be borne out with a remark-
able degree of regularity. Figure 5.3 shows the patterns of total consumption which each of the six countries generated over the decade.
Australia is particularly interesting as the effects of its first mineral
boom caused consumption to grow very quickly up to 1974/5 when the

Figure 5.4: Production in Manufacturing Industries (1970 = 100)

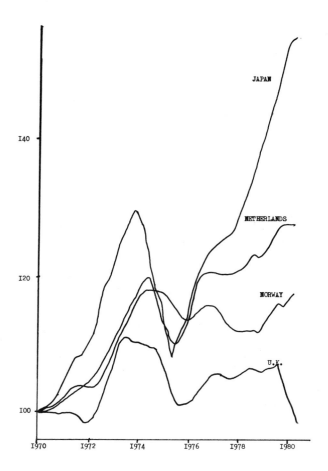

Source: OECD Indicators of Industrial Activity.

oil price rise neutralised its effects and consumption fell to a much lower growth path. The overall behaviour of the Japanese economy can also be quickly grasped from this figure. From 1970-3 Japanese consumption was growing at a very fast rate; between 1974 and 1976 this growth trend was slower as the economy adjusted to the new oil policies, then growth continued almost unaffected. The contrast between the relative fall in consumption in Japan, Australia and the UK and the rise in Norway and the Netherlands, shows the redistribution in real wealth which the oil price rise represented.

The cases of Australia, Japan and Venezuela are of interest in a general way as they throw light on the way that the Gregory thesis works in different economies. The cases of Norway and the Netherlands are of much greater interest to the UK, however, as they represent two of the extremes of policy which might be adopted in this country. Figure 5.3 shows that on the basis of *per capita* consumption the Netherlands' policy seems to be preferable. In 1970 Norway had the higher level of *per capita* consumption; by 1980 the position was reversed.

Figure 5.4 shows how industrial production in manufacturing industry has changed over the same period. It is clear that since 1975 the Netherlands industrial sector has performed significantly better than the Norwegian sector. This is a surprising conclusion as it is contrary to the popular view. But it does seem that industrial production may be more effectively promoted by stimulating future demand and allowing weak industries to fail rather than by trying to maintain the existing industrial structure unchanged. The revenue generated by the natural resource should therefore be channelled first only into those areas of the domestic economy which can offer a realistic investment prospect, and secondly to relieve the social hardship which will inevitably result from structural change. To aim simply to maintain employment levels may lead to investment being channelled into those industries which are most inefficient and which are least able to provide any long-term return on the investment.

Part III

OIL AND THE ECONOMY

6 OIL AND ECONOMIC POLICY

The development of the oil and gas sector affects the economy in three main ways. There are the direct effects, there are effects on industry and there is the policy response of the government. The direct effects have been discussed in Part I of this book. They consist of such things as the contribution of oil and gas to the gross domestic product, the balance of payments effect, and the direct employment effect. The immediate changes in industrial structure flowing from the development of oil have been considered in Part II of this book. These are largely brought about through changes in the exchange rate and involve shifts in the relative size of various productive sectors as well as changing employment patterns and increased consumption levels. The third channel through which North Sea oil affects the economy is the main concern of this part of the book. It consists of the response by the government in terms of its general economic policy. The arrival of oil production on a big scale makes it easier for a government to follow expansionary policies, if that is desired, because the oil greatly improves the balance of payments and supports the exchange rate. If, however, the government wishes to follow a restrictive economic policy aimed at curbing inflation, it will welcome and not offset the appreciation of the exchange rate which results from oil production, and will use the revenues from oil taxation to tighten its fiscal stance rather than to expand demand.

The Relationship Between North Sea Oil, the Exchange Rate and the Economy

Before examining government policy, we look at the exchange rate mechanism. This important mechanism through which the oil sector affects the rest of the economy has two links. The first is the determination of the exchange rate itself; the second is the effect of the exchange rate on the macroeconomy. Of these two links the latter is more clearly understood than the former. No one has yet succeeded in modelling the exchange rate adequately, and this is one of the

127

weakest areas of all the large macroeconomic models. Haache and Townend[1] have recently reported on a major piece of research on the exchange rate and the first lines of their conclusion summarise the present situation. 'The predominant impression left by our results is one of failure. We have not succeeded in finding empirical regularities in the data to help explain in any satisfactory way the fundamental determinants of sterling's effective exchange rate'. Yet the exchange rate is a fairly simple concept in economic theory.

At the simplest level the exchange rate is simply a price which moves to clear the foreign exchange markets. So, as a simple example, if one hundred pounds are brought to the market to exchange for dollars, and two hundred dollars are brought to exchange for pounds, then a rate of two dollars to the pound will clear the market. That is, everyone who wants to change pounds for dollars, or dollars for pounds, is satisfied. If only one hundred and fifty dollars were presented for exchange into sterling then the exchange rate would fall to 1.50 dollars to the pound. This is the simple process by which the exchange rate is determined; the complications arise when we try to explain the flows of sterling and foreign currencies on to the market.

One of the most common approaches to this problem has been to take the main determinant of the flow of currency on the foreign exchange markets to be the level of imports and exports of goods and services. The level of British imports represents the demand to exchange sterling for another currency, and British exports represent the demand to exchange other currencies for sterling. This is the basis of exchange rate determination used by Gregory and outlined in Chapter 4. The exchange rate is assumed to move to maintain equilibrium on the current account of the balance of payments. When there is a surplus the exchange rate rises and when there is a deficit it falls. According to this view, North Sea oil would have large effects on the exchange rate, since it contributes very large amounts in terms of exports and import saving to the balance of payments.

There are, however, two other ways of explaining the exchange rate, both of which assign a much less important role to North Sea oil. The first of these is the Purchasing Power Parity theory of the exchange rate. This is a very longstanding proposition in international economics. It says that the exchange rate will move so that the price of goods will be equal across international boundaries. It is sometimes referred to as the law of one price. If a car can be purchased for £5,000 in England and $10,000 in America, then the exchange rate must be $2 to the pound. If the exchange rate were to be only $1 to the pound, then

Americans would import cars from Britain, a balance of trade surplus would result and the exchange rate would rise. This process would continue until the exchange rate reached $2. Similarly, there will be a close link between the movement of prices in a country and its exchange rate. If domestic prices double while prices remain unchanged in the rest of the world, then the exchange rate must fall to half its original value according to this theory.

These last two propositions are, however, slightly different from one another. It is widely accepted that a relative increase in prices will lead to a proportionately inverse fall in the exchange rate, although many economists might qualify this by noting a fairly considerable time lag in the relationship. But the idea that goods must sell at one price internationally is subject to qualifications. The first qualification is due to transport costs; the second is that goods from different countries may be inherently different. A French car is not the same thing as a Japanese car. The theory is based on the assumption that the world demand for a country's goods is infinitely elastic at the ruling world price. If this is not the case, then the exchange rate can change permanently over time. If, however, this assumption were valid then the exchange rate could not be permanently affected by North Sea oil.

The third main approach to the determination of currency flows on the foreign exchange markets concentrates primarily on the capital account of the balance of payments. In this view, the main determinant of the exchange rate, at least in the short run, is the movement of money between the domestic and foreign money and capital markets. The exchange rate is fixed so as to bring asset markets into equilibrium (and the current account is only required to balance in the long run). On this view the exchange rate can vary substantially over a short period of time as asset markets are disturbed by interest rate movements. This last factor is particularly important as it weakens the links between current oil production and the exchange rate. As soon as the oil is discovered the exchange rate should rise because of the expectation that the exchange rate will rise in the future. This means that the actual level of oil production may be less important than the fact that the oil exists.

These three basic views of the determination of the exchange rate are not mutually exclusive. It can easily be argued that the capital account approach determines short-term fluctuations in the exchange rate while the trade balance determines the long-run exchange rate corrected for price movements and Purchasing Power Parity determines the nominal exchange rate in the longer run. It is clear, however, that it is not easy to determine the force of the various factors that bear on the exchange rate. While North Sea oil should affect the exchange rate under anything but the most extreme assumptions, the size and timing of its impact are highly uncertain.

Figure 6.1 shows the paths of the effective sterling exchange rate, relative export prices and short-term interest rates. The relative export prices trend is an index of the real exchange rate, that is the exchange rate adjusted for world and domestic inflation rates. This removes the Purchasing Power Parity effect from the exchange rate. It is clear that high UK inflation rates have resulted in a long-term decline in the nominal exchange rate. The interest rate-real exchange rate relationship is interesting; before 1978 the two series seem to be negatively correlated, while after 1978 there seems to be a fairly good positive association between them. The explanation for this change is that while in the early part of the period the government was trying to influence the exchange rate by controlling interest rates, in the latter part interest rates were set for domestic reasons. So when the exchange rate began to fall in 1972, interest rates were raised to prevent the fall. From 1976, as the exchange rate began to appreciate, interest rates were reduced in an attempt to control the appreciation. During the last part of the period, however, government policy towards the exchange rate has changed. Recently interest rates have been set according to domestic considerations, in particular the money supply, and so it may be that the positive relationship between interest rates and the exchange rate which theory suggests is more apparent.

What is not clear from this table is how important North Sea oil has been to the real exchange rate changes. There are two ways in which the oil could have affected this graph; first during the period 1973/4/5 the expectation of the oil could have prevented a sharp decline in the real exchange rate which might have resulted from the 1973/4 oil price rise; and secondly during the period 1977-80 it could have caused an increase in the real exchange rate as the rate of oil production increased. Both of these arguments are plausible and fit the facts. The problem is that the period 1973-5 was a high interest rate one and the period 1977-80 was one of dramatically rising interest rates. This explanation fits the facts equally well. So it is not a question of one explanation or the other, but rather how important each of the effects has been.

As long as the two possible effects work together there can be no definitive answer to this question. As an example of the confusion which exists over this point, a recent article by Beenstock, Budd and

Figure 6.1: Interest Rates and the Exchange Rate

a = relative export prices
b = sterling effective exchange rate
c = treasury bill yield

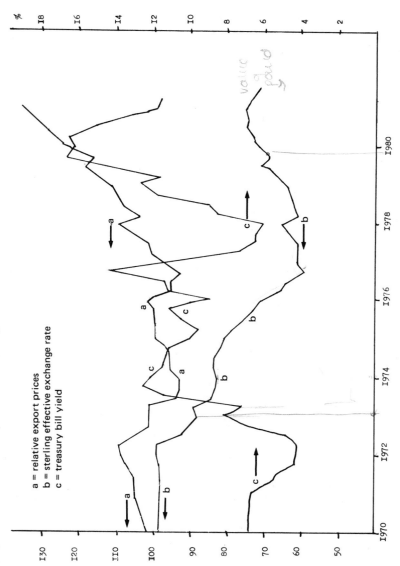

Warburton[2] estimated a number of single equation exchange rate models and concluded that North Sea oil was responsible for a 14 per cent increase in the real exchange rate. On the other hand Haache and Townend[3] achieve a better fitting equation which does not even include North Sea oil. Both equations are unable to forecast outside the estimation periods, and are therefore unsatisfactory.

Having outlined the difficulty of apportioning responsibility for the exchange rate rise between interest rates and North Sea oil, it is still necessary to make some sort of judgement in order to assess the effect of oil on the economy. From 1977 to 1980, the real exchange rate rose by about 35 per cent while the nominal exchange rate rose by about 25 per cent. It seems likely, particularly in view of the recent fall in exchange rates, that the primary responsibility for this rise was interest rate movements. On this basis we might assess the effect of oil on the real exchange rate during this period as about 10 per cent.[4] It may be argued that an extra amount should be added to this because of the absence of a fall in the real exchange rate in 1974 and 1975, which may have been due to North Sea expectations: this can hardly be quantified.

The impact of the exchange rate on the rest of the economy is fortunately quite well understood. Both output and prices are affected. An increase in the exchange rate means that the sterling price of all imports goes down; it also means that the foreign exchange price of all our exports is increased. This change in the price of imports will feed through into the whole economy, lowering the price of all goods which are affected by these imports. Ultimately an overall reduction in domestic prices would result. Further, the purchasing power parity theory suggests that over time the fall in domestic prices would be proportionate to the rise in the exchange rate. In fact many of the large economic models do find that in the medium to long run, five years or more, the domestic price level moves closely in accord with the purchasing power parity theory.

Before the price effects have completely worked through the economy, a number of other changes will take place. The immediate impact of a sudden increase in the exchange rate will be an actual improvement in the current balance of payments. This comes about primarily because the actual quantities of goods traded cannot adjust immediately and so the change in prices (in sterling terms, cheaper imports while export prices remain unchanged) produces a short-term surplus on the balance of payments. As time passes the actual quantities of exports and imports will adjust to the new price relationships. That is, the demand for our exports will fall, due to their higher price in terms of foreign currency and the quantity of goods imported will rise due to the lower sterling price of imports. These effects will combine to produce a worsening of the current account of the balance of payments and an overall reduction in domestic production, due partly to a decline in exports and partly to imported goods being substituted for domestically produced goods. These effects on output will tend to disappear as domestic prices adjust and the real rate of exchange eventually returns to its former level.

Table 6.1 presents a summary of the predicted effects of a 5 per cent appreciation in the exchange rate given by three of the large economic models. The adjustment of prices and wages to the 5 per cent exchange rate appreciation is shown in the first three columns. The London Business School shows a slightly faster rate of both wholesale price adjustment and average earnings adjustment than NIESR and the Treasury model. This faster price adjustment results in a much weaker real effect. The GDP change in the fourth column is negligible for the LBS model while being significantly negative in the first three years in the National Institute's simulation. The current balance behaves much as predicted by theory with a significant positive J curve effect in the first year followed by a series of negative effects.

According to these estimates the appreciation of the effective sterling exchange rate between 1978 and 1980 of 25 per cent would have produced a price level between 8.5 per cent and 14.5 per cent lower in 1981 than it would have been in the absence of the appreciation (depending on which model estimate is used). The rise in the exchange rate would also have caused a 2 per cent drop in real GDP in 1980, and a 1 per cent drop in real GDP in 1981. It is, of course, necessary to stress that as oil was only partly responsible for the exchange rate appreciation, it is only partly responsible for these effects on prices and output.

An additional complication which must be dealt with when the effects of a change in competitiveness are being considered is the structural shifts which may have occurred within individual parts of the economy. This was considered, in a general way, in Chapter 4. A change in the real exchange rate will have different effects on individual sectors of the economy. In Chapter 4 a general division into manufactured goods (which were taken to be tradeables) and other goods (which were taken to be non-tradeables) was made. This is, of course, an oversimplification: virtually all goods are traded to some extent, so we might expect the goods which are more heavily traded to be more affected by a change in the real exchange rate but all sectors will experience some effects.

Table 6.1: Effects of 5 per cent Appreciation in Exchange Rate

End Year	Wholesale prices (%)			Competitiveness (%)			Average earnings (%)		Real GDP (%)		Current balance (£b)	
	HMT	LBS	NIESR	HMT	LBS	NIESR	LBS[a]	NIESR	LBS	NIESR	LBS[b]	NIESR
1.	−.9	−2.0	−1.1	−1.7	−3.0	−3.3	−0.9	−0.1	+0.1	−0.4	+0.7	+0.3
2.	−1.7	−2.9	−2.7	−1.7	−1.9	−1.6	−2.1	−1.1	+0.1	−0.2	−0.8	−0.1
3.	−2.6	−3.6	−3.2	−1.2	−1.2	−1.3	−3.4	−1.9	−0.2	−0.1	−1.2	−0.2
4.	−3.1	−4.3	−3.5	−0.9	−0.4	−0.9	−4.5	−2.6	−0.1	+0.1	−1.3	−0.4
5.	−3.6	−4.9	−3.8	−0.7	+0.2	−0.6	−5.4	−3.0	−0.1	+0.2	−1.2	−0.4

Notes: a. Manufacturing only.
b. Balance of Official Financing.
Source: *Britain's Trade and Exchange Rate Policy*, ed. Robin Major, HEB 1979, pp. 113 and 127; and HM Treasury, 1979.

In order to gain some idea of the magnitude of these effects we need to use an economic model which contains a detailed breakdown of the industrial sector. The Treasury carried out an exercise of this nature using the Cambridge[5] Econometrics model which is particularly suitable for this type of analysis. Their results are reported in Table 6.2. This simulation shows the relative effect on 31 industries of a 10 per cent appreciation in the real exchange rate over a three-year and a 10-year period. The industries are classified into one of nine groups depending on whether the size of the decline in output over each of the two time horizons is marked, moderate or minimal. (Over the three-year period a minimal effect is less than ¾ per cent change in output, a moderate effect is ¾ per cent to 1½ per cent and a marked effect is more than 1 per cent.)

The broad analysis of Chapter 4 is confirmed in this table. Those groups which experience the strongest effect from the appreciation tend to be the groups which are most obviously in the traded goods sectors. The largest effects occur in the vehicles and electrical and instrument engineering industries, while service sectors such as transport, water, etc., show only a minimal response. This analysis also supports the suggestion in Chapter 4 that the sectoral responses to exchange rate changes take a very long time. The differences between the pattern of changes after three and ten years is really quite marked. So a lot of the adjustment is occurring between these years.

North Sea Oil, Government Policy and the Economy

This section deals with the important link between North Sea oil and the macroeconomy via the tax revenues generated in the oil sector and the use to which the government puts them. The way in which this link operates depends on how the government treats the oil revenue: whether it goes into general tax revenues or a special 'oil' fund; and more importantly whether the government chooses to spend the oil revenues or to use them to reduce the budget deficit. This latter key decision will reflect a government's basic economic policy.

The Treatment of Oil Tax Revenues

It has often been suggested that a separate 'oil' fund should be set up with the tax revenues generated by the North Sea sector.[6] The basic argument for doing this is that it would allow the oil revenues to be easily identified and that they would then be put to productive use rather than being 'frittered away' as part of general government

Table 6.2: Response of non-oil Output to Changes in Exchange Rate (figures in brackets are proportions of GDP, and these are totalled in the boxes[a])

Effect after 10 years \ Effect after 3 years	Marked	Moderate	Minimal
Minimal		Mining (2.0) Chemicals (2.5) Bricks, etc (1.1) Timber and furniture (1.0) Printing & publishing (2.2) Construction (6.3) Gas & Electricity (2.7) Communications (2.7) Distribution (10.6) Financial Services (8.3) Professional Services (8.8) 48	Agriculture (2.5) Food, Drink & Tobacco (3.4) Mineral oil refining (0.4) Aerospace equipment (0.7) Water (0.4) Transport (5.6) Miscellaneous Services (10.7) Government and Defence (7.0) 31

Moderate	Iron & Steel	(1.2)
	Mechanical Engineering	(3.9)
	Metal Goods	(1.8)
	Textiles	(1.4)
	Leather & Clothing	(1.0)
	Paper & Board	(0.3)
	Manufacturing	(1.3)
	11	
Marked	Instrument engineering	(0.6)
	Electrical engineering	(2.7)
	Vehicles	(3.1)
	6	
	Non-ferrous metals	(0.5)
	Shipbuilding	(0.5)
	1	

Note: a. Excluding 'petroleum and natural gas' and 'ownership of dwellings'.
Source: Treasury Working Paper, No. 22.

consumption. This fund would allow certain projects to be clearly defined as North Sea funded and so the benefits of the oil revenues would be clearly apparent. No British government of either party has accepted this view, however.

The opposite view is that all the government's tax receipts should be pooled: the revenues generated in the North Sea should be handled in just the same way as any other source of government revenue. It is also argued that it would be wrong to undertake an investment project simply because the funds for it come from the North Sea rather than from some other source. Any investment project should stand or fall on its own merits rather than on the origin of the money funding it. North Sea oil revenues should therefore be added to general government revenues and allocated in precisely the same way as any other tax revenue.

In our view there is no great virtue in creating a special fund from North Sea oil revenues. Nonetheless the nature of the revenues, flowing from an exhaustible natural resource, does suggest that some of the revenues should be used to finance additional investment which would leave the country with a better capital stock when the oil runs out. Equally, it is perfectly legitimate, as maintained in Chapter 3, to enjoy some of the benefits of North Sea oil and gas in current consumption. This points to tax reductions. These considerations are pursued in the final chapter on alternative policies.

Government Economic Policy

The present government has employed the benefits of North Sea oil to further its restrictive economic policies. What is the rationale? The traditional approach of postwar governments has been to aim at four goals: economic growth, stable prices, full employment and a satisfactory balance of payments. A surplus on the current balance of payments is not desirable in its own right, but it is necessary that a long-term deficit be avoided for a generally healthy economy. Stable prices have traditionally been seen as a less important objective: the aggregate damage to an economy of a moderate and fairly stable inflation rate was generally considered to be quite small in the 1950s and 1960s. The two main indicators of welfare were traditionally full employment, which indicates maximum production in the short term, and economic growth, which indicates maximum production in the long term. Of these two aims full employment, as the short-term indicator of welfare, has been the prime objective of governments from World War II until fairly recently.

The rise in the rate of inflation over the last 10 to 15 years has caused more emphasis to be placed on the goal of stable prices, but until recently full employment remained the prime objective of economic policy. This changed with the election in May 1979 of the present government whose primary objective, in the short term, is to reduce the rate of inflation. This does not necessarily represent an overall re-ordering of objectives; in the long run full employment and economic growth are still the ultimate objectives. But the present government believes that these can only be achieved if inflation is first brought down to a lower level. So, given this view of the economy, the sole objective of economic policy becomes the control of inflation. The other three traditional objectives are largely ignored.

This switch in objectives is based primarily on the government's view of how the economy works. If its view is accepted then there can be little dispute that price stability should be the exclusive short-term objective of the government. If current inflation will really prevent future growth and cause unemployment then of course it must be stopped as a first priority. However, the experience of high inflation together with high growth in many countries, over a long period, suggests that this view is mistaken.

To understand the government's attitude it is necessary to recall the essential features of the monetarist thinking which Ministers came to adopt while in opposition. To begin with there are some monetarist propositions which can be readily accepted. For example, the statement that inflation is a monetary phenomenon can be accepted by many non-monetarist economists. Monetarists have put considerable effort into proving its validity on both a theoretical and an empirical basis. Prices cannot rise in a continuous fashion unless the money supply also rises. Similarly, it is undisputed that an autonomous and sustained increase in the money supply will eventually cause prices to rise. Where the monetarist doctrine becomes questionable is in developing three further propositions.

1. The money supply is under the control of the government and the direction of causation runs from money to prices.
2. There is a natural rate of unemployment and output and the economy will adjust to this level reasonably quickly.
3. An increase in the public sector borrowing requirement and the national debt will ultimately lead to an increase in the money supply.

Each of these propositions will be considered in turn.

While it is true that inflation is a monetary phenomenon this does not mean that there is a single direction of causality from money to prices, nor that the money supply is actually under the control of the government. The question of whether or not the money supply can be controlled raises the subsidiary problem of the correct definition of money. There are at least seven possible contenders for the title of money (the monetary base, Retail M1, M1, Sterling M3, M3, Private sector liquidity 1, and Private sector liquidity 2). It is often argued in theoretical work that this is an unimportant point as the various definitions of money will move in a uniform way. This is not true, however, as Figure 6.2 shows: the relationship between the various money stock definitions in the medium term (e.g. 5 years) is very weak indeed.

There is similar confusion over the definition of inflation and the price level; it is possible to construct a price index in a number of varying ways, i.e. the Consumer Price Index, Retail Price Index, the GDP deflator, Wholesale Price Index, etc. Once again, monetarist theory often assumes that the differences are unimportant and that all the indices move together, but again this is not necessarily so. Figure 6.3 shows the rates of change of the retail price index, wholesale output prices and wholesale input prices; there is a clear divergence between them. The choice of the appropriate definition of both money and prices is therefore both difficult and very important. In practice this choice is generally made on the basis of which series best explains inflation over any given period. But with so many series available it is almost inevitable that one or another will correlate well with inflation. This does not mean that the correlation can be expected to hold in the future.

It is also quite clear from Figure 6.2 that the government has been unable to control the rate of growth of M3. This is made up largely of bank deposits which are determined by the interaction of the banking sector and the demand for money from the public at large. The government has no direct way of controlling this definition of money. Its influence over the demand for money is via interest rates and this may be unreliable.

The question of whether money is demand or supply determined is a very important one for monetarism. If the supply of money can be controlled then the long-run association between money and prices, which has been conceded, will ultimately allow the control of inflation. If, however, money is demand determined then inflation will be an important determinant of that demand. So if inflation is high the

Figure 6.2: The Rate of Growth of the Monetary Based, M1 and Sterling M3

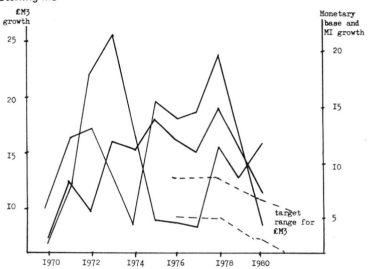

Source: Bank of England Quarterly *Bulletin*.

Figure 6.3: The Rate of Change in Three Price Indices

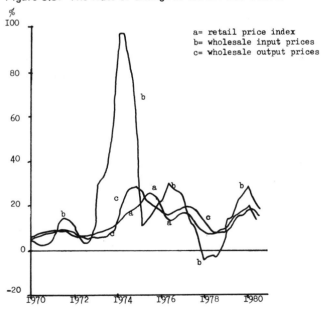

level of interest rates necessary to reduce the demand for money to meet some arbitrary target may be quite unacceptable.

Professor Friedman, in his evidence to the Treasury and Civil Service Select Committee,[7] made the point very forcibly that controlling the quantity of money by manipulating interest rates and the demand for money was not monetarism as he believed in it. He suggests that the whole British financial system should be restructured so that the supply of money can be controlled by manipulating the monetary base. It is not possible to say if this approach would prove practical but it is certainly possible to assert that the present system makes it impossible to control the money supply. The demand for money can be influenced over the medium term, but the effects of such action cannot be linked unambiguously with the price level, nor analysed in the terms of a simple monetarist framework.

The second proposition, that the economy will move quickly towards its natural rate of output and growth, is completely separate from the first. It is quite conceivable that the one could be true while the other is false. The reason that they both form a part of monetarist doctrines is partly that the early quantity theory of money formulation used the notion of long-run full employment and partly that Friedman developed the concept of the natural rate of unemployment as a counter to the Phillips curve proposition of a long-run influence of unemployment on prices. If the price level is determined solely by the quantity of money, in the long run it is not possible to have a persisting trade-off between unemployment and the rate of inflation. The argument for a trade-off between inflation and unemployment had therefore to be refuted if monetarism were to be accepted. Friedman attempted this by developing the concept of a natural rate of unemployment.

When monetarist doctrines are actually put into practice this proposition becomes very important. If we ignore the timing, then most monetarists and Keynesians would agree on the broad order of events following the implementation of a restrictive monetary policy. That is, the government's attempt to reduce the rate of growth of the money supply, whether successful or not, will tend to raise interest rates and the exchange rate. The rate of growth of output will fall, partly because of a fall in investment, especially investment in stocks, and partly because of greater competition from foreign producers. This will give rise to an increase in unemployment. After a time the rate of inflation will begin to fall either because of the high exchange rate or the fall in domestic demand, or both. As price inflation falls the output effects will diminish, the exchange rate will return to its former level and unemployment will fall. A major division between the monetarists and Keynesians is over the length of time this might take. A typical Keynesian view would be that the initial fall in real output would occur fairly quickly, but that the subsequent effect on prices could be delayed for a number of years and that even when inflation fell there would be no strong force operating automatically to bring real output back to the full employment level. Many supporters of monetarist policies would give a much shorter estimate of the time needed for the economy to adjust to the policy shock.

The speed of adjustment of an economy to such a monetary shock is essentially an empirical question. Table 6.1 shows that three of the major economic models show significant deviations from a base trend even after five years, resulting from quite a small shock, and the other large models would show similar results. When examining single equation tests of the relationship between money and prices and the speed of adjustment it is easy to become quickly swamped by the number of sets of results available. We have therefore picked out one set of results only but these are produced by a research group sympathetic to monetarist views. In September 1981 a group[8] from the London Business School reported the following equation:

$$P_t = .1715M_{t-5} + .1930M_{t-5} + .1633M_{-1} + .1685M_{t-8}$$
$$+ .0253M_{t-9} + .0195M_{t-10} - .0020M_{t-11} - .0064M_{t-12}$$
$$+ .1587M_{t-13} + .1900M_{t-14} + .1539M_{t-15} + e_t - .0041$$
$$P = \triangle \text{ Log wholesale price index}$$
$$M = \triangle \text{ Log sterling M3}$$

Figure 6.4 shows the adjustment path of inflation given by this equation if a steady rate of money growth of 10 per cent is reduced to 5 per cent. There is a delay of over a year before any effect occurs and the full reduction in prices would not result until nearly four years later. Throughout all this period real output is depressed and the recovery does not even begin until this period is over. This is an estimate which is sympathetic to the monetarist view. The Keynesian view might well be that there would be little or no recovery of output after the rate of inflation has stabilised.

The question of the stability of the economy and how quickly it adjusts to any shocks can be approached in a number of ways. A casual empirical look at the real world shows economies with high

Figure 6.4: The Adjustment of Inflation to a Change in the Rate of Growth of Money

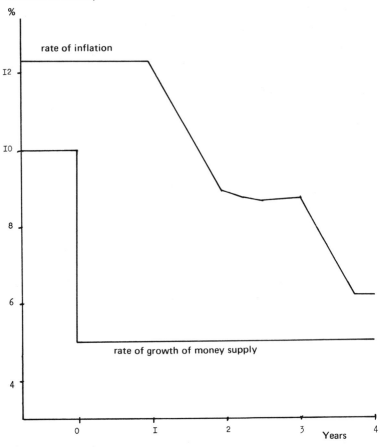

unemployment for a considerable period of time that do not seem to have a strong tendency to return to a full employment equilibrium. Secondly, a study of the large macroeconomic models shows that they can remain out of equilibrium for a considerable time and most of the models do not exhibit any long-run tendency towards full employment.

Finally, the estimates made by monetarists themselves seem to represent a very long-run adjustment process in terms of normal economic policy.

The final proposition, that an increase in the national debt will lead to a future increase in the money supply, is perhaps the most important of the three. This is not because it is a very necessary part of monetarism, but because it is responsible for the main area of disagreement between monetarists and non-monetarists. Many non-monetarists might be happy to accept a 'sound' monetary policy in the sense of trying to restrict the rate of growth of the money supply to a rate slightly below the rate of inflation plus a natural rate of growth of output. What they do not accept is the reduction in government spending and the overall restrictive fiscal policy which the present government has adopted in its attempts to control the money supply. If this third proposition is ignored, for the moment, it is easy to see that it is quite possible for the government to reduce the rate of growth of the money supply without cutting its spending. The government has to operate within the confines of the following budget constraint:

$$G = T + \triangle M + \triangle B - rB$$

That is, government expenditure (G) is equal to total taxation (T) plus the new issues of money ($\triangle M$) plus the increase in government borrowing (B) less the interest paid on outstanding government debt. Given this constraint it is easy to see that the money supply can be controlled without affecting government expenditure or taxation simply by increasing government borrowing ($\triangle B$).

If the final proposition is accepted and an increase in public debt implies an increase in the money supply, then the only way the money supply can be controlled is by increasing taxation or reducing government expenditure. So the final proposition, if it is accepted, implies a deflationary fiscal policy which is unacceptable to many non-monetarists.

Before the arguments in support of this proposition are examined it is necessary to stress that many monetarists would be unwilling to accept the validity or consequences of it. Professor D.E.W. Laidler has stated that 'I do not believe that the links in the British economy between the balance of the budget and the Public Sector Borrowing Requirement on the one hand and the rate of money creation on the other are so clear cut and rigid that the adoption of monetary targets has any strong implication for the conduct of fiscal policy.'[9] Similarly Professor Friedman, when commenting on the current policies, said 'The key role assigned to targets for the PSBR, on the other hand, seems to me unwise for several reasons. (1) These numbers are highly misleading because of the failure to adjust for the effect of inflation. (2) There is no necessary relation between the size of PSBR and monetary growth'.[10] This proposition cannot therefore be presented as a general monetarist one but rather as an idea specific to the ideology of the present UK administration.

The government's view of the importance of this proposition is illustrated in the following quotation.

Experience shows the difficulty of attempting to finance a series of excessive public sector deficits without adding to the money supply. Excess budget deficits increase the supply of financial assets in the economy. If the total quantity of financial assets is growing rapidly it is only possible to keep down the growth of the money supply by higher and higher interest rates.[11]

This statement supports the proposition in two quite different ways. Firstly, it suggests that experience confirms the proposition which is empirically provable. Secondly, it suggests a theoretical linkage between the quantity of money, government debt and interest rates such that an increase in government debt will lead to an increase in the quantity of money.

It is helpful to specify this relationship more fully. There are two distinct links in the Treasury argument. The first is between the PSBR and total wealth: an increase in the PSBR is said to cause an increase in total nominal wealth. The second linkage is between total wealth and the money supply through the demand for money. It is accepted that the quantity of money is demand determined and the demand for money is taken to be a stable function of interest rates and total nominal wealth. If interest rates rise the demand for money will decline while if total wealth rises the demand for money will expand. Given this fairly simple money demand function, any increase in wealth must cause a rise either in interest rates or in the quantity of money. If the government is not prepared to accept rising interest rates and if it wishes to reduce the quantity of money demanded, a reduction must be made in total wealth. The first linkage, between the PSBR and total wealth, suggests that this should be done by reducing the PSBR.

The first thing which is made clear by this account is that the Treasury is not trying to control the money supply, but rather the demand for money. This is certainly not in accordance with the views of traditional monetarists such as Friedman. The second point is that the general theoretical mechanism is open to criticism at almost every step. First the primary linkage between the PSBR and total wealth is very weak; there is even considerable doubt as to whether government bonds should be counted as net wealth (as they represent a stream of future tax liabilities). The Treasury has itself spent considerable effort trying to find evidence to support this link and has concluded that 'In practice no simple empirical relationship between movements in gross nominal wealth and in the public sector financial deficit was found.'[12] The link between total wealth, money and interest rates, through a demand for money function, is also weak. Evidence for a stable money demand function is slight in general, and the assumption that this function will contain nominal financial wealth rather than income or the value of total real assets has little support in fact. The Treasury has provided some empirical backing for this specific function in its evidence to the Select Committee on Monetary Policy,[13] but this evidence is unconvincing. Professor D. Hendry, a special adviser to the Committee, commented on the Treasury evidence: 'These results do not seem to be credible from any basis I can imagine.'[14] There is no theoretical reason to expect the Treasury function to be the correct one, while the empirical evidence in its favour is negligible.

The theoretical arguments for a strong link between the PSBR and the money supply are thus doubtful. But the Treasury statement did not rely solely on such abstract ideas. It claimed that 'experience' provides the necessary confirmation. At the empirical level the argument can be supported either by showing that there is a stable long-run relationship between the nominal national debt and the money supply, or by estimating an equation for the movement of the money supply and using the PSBR as one of the explanatory variables. A good example of this last approach is the work of Beenstock and Longbottom,[15] where an equation for money supply movements is estimated and the PSBR is found to be a significant variable.

It is, of course, inevitable that there will be a close association between the national debt and the money supply over a long period. This is simply because both are nominal variables so they grow over time at a similar rate. This does not mean that an increase in the PSBR has a causal effect on the money supply. The more formal econometric evidence of this link is very weak. Cuthbertson, Henry, Mayes and Savage[16] have shown that the Beenstock/Longbottom model is not stable and that even using the original data a slightly better (in terms of R^2) equation can be obtained simply by omitting the PSBR from the equation completely.

What is a Deflationary Budget?

This is an important question. If government expenditure is greater than taxation, how can the budget be described as deflationary? In conventional textbook economics a budget is defined as deflationary only when government expenditure is less than total taxation. When total government expenditure is greater than total taxation the budget is defined as expansionary. However, the simple definition of the state of a budget applies only to an economy which is neither growing nor experiencing inflation. Inflation represents a very large and important tax on the national debt holders which is not included in the conventional taxation account.[17] If the private sector holds £100,000 million of national debt, approximately the level in 1980, and inflation takes place at 15 per cent, for the real value of the debt assets to remain unchanged they should be valued at £115,000 million. This will not happen. As a first approximation their nominal value will be constant (although if interest rates rise the value may actually fall) so in effect £15,000 million has been transferred from the private sector to the government by the 15 per cent inflation. If the government runs a conventional deficit of £15,000 million it will only be replacing the hidden taxation and a neutral budget will be achieved.

In a time of inflation it is not correct to look at the difference between taxation and expenditure to gauge whether a budget is inflationary or deflationary. There is probably no ideal measure of this, but the movement of the national debt as a percentage of GDP is probably a better guide than any other (this is shown in Figure 6.5). On this basis there is a need for a considerable expansion of the public sector deficit. For example, merely to return to the national debt relationship of 1970 would entail an increase in the national debt of £50,000 million. It would not be desirable to do this in one year but over a number of years it may be a desirable aim. It is surprising that 45 years after the publication of Keynes' general theory many people feel that a budget deficit is in principle wrong, that it is something to be ashamed of. This fact alone is sufficient reason to focus attention on the national debt-GDP relationship rather than the PSBR.

There is another concept of the budget balance which has received wide attention over the last 20 years. This is the full-employment or standardised balanced budget. This concept recognises that both government expenditure and total tax revenues will vary as the level of economic activity changes. Government expenditure will tend to fall as output rises because of a fall in unemployment and a consequent

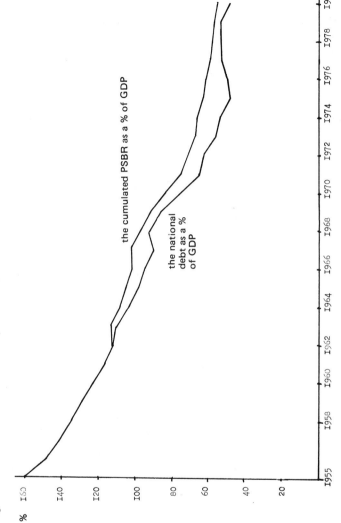

Figure 6.5: The National Debt as a % of GDP, 1955-80

the cumulated PSBR as a % of GDP

the national debt as a % of GDP

Source: Annual Abstract.

fall in the cost of social security and other benefits. Total taxation will tend to rise as a result of the proportional taxes on income, expenditure and profits. So as national income rises the need for government borrowing declines.

This concept of the standardised budget is not at variance with the inflation adjusted proposal given above. The standardised budget approach does not take account of inflation and so it suffers from the same defects as the conventional approach. It would be quite valid to combine the inflation adjusted approach and the standardised approach. So, at the desired level of employment, the GDP-national debt relationship should remain constant. If the economy is operating at employment levels below this desired level the GDP-national debt relationship should rise, and if it is above this level it should fall.

Table 6.3: The Standardised Budget, 1974/5-1981/2

	74/5	75/6	76/7	77/8	78/9	79/80	80/81	81/82
		PSBR as a % of GDP on a standardised basis						
Change in fiscal stance		−1.97	−1.94	−2.23	1.06	−1.32	−2.12	−3.29
Level of fiscal stance	9.5	7.5	5.6	3.36	4.42	3.1	.89	−2.4

Source: National Institute Economic Review, 1982/1; Savage, pp. 85-96.

Table 6.3 gives some estimates for the movement of the standardised budget from 1974-5 onwards. When trying to assess figures for the standardised budget one of the main difficulties is the choice of actual level of output which should be used in the comparison. When the concept of the standardised budget was originally proposed the obvious level of output was the full employment one. It is now almost impossible to define the full employment level of output so some other base must be used. Table 6.3 uses the level of output and employment in 1974-5 as its base; unemployment was around 2.5 per cent. In this year the standardised deficit is therefore equal to the actual deficit. Subsequent figures show the budgetary stance on the assumption that the unemployment levels are maintained. The table shows both the change and the level of the standardised budget in each year. It is apparent that fiscal policy became increasingly restrictive and that its effect has become negative under the present administration.

Conclusion

It is undeniable that inflation is a monetary phenomenon and that a sound monetary policy should be an objective of any government. This does not mean that the money supply can be easily defined or controlled, nor does it mean that monetary policy will work either quickly or with little disturbance to the economy. Our view is that the money supply is both hard to define and control and that the adjustment to any monetary change will be long and will involve widespread real disturbance. We do not accept the view that there is a causal link between the PSBR and the money supply, and so we see no need for the highly deflationary fiscal policies which have been adopted since 1979.

We take a broadly Keynesian view of the working of the economy. That is to say, we do not believe that the economy exhibits any strong automatic tendency to move towards a high level of economic activity with low unemployment within a reasonable time horizon. We accept that high interest rates, reduced government expenditure, high taxation, and attempts to restrict monetary growth will ultimately reduce inflation, but we believe that this works only by creating a substantial and lasting economic depression. The cost of the depression, in our view, outweighs the gains in terms of reduced inflation. Unless more rational and direct methods of curbing inflation are accepted as practicable, some use will have to be made of restrictive fiscal and monetary policies and considerable unemployment accepted. But it does not make sense to abandon the aim of high employment and economic growth. A reasonable compromise should be sought.

Notes

1. G. Haache and J. Townend, 'Exchange rates and monetary policy: modelling sterling's effective exchange rate 1972-80', *Oxford Economic Papers*, No. 33, Special Supplement. This piece of work was carried out at the Bank of England.

2. The authors were at the London Business School. M. Beenstock, A. Budd, and P. Warburton, 'Monetary Policy Expectations and Real Exchange Rate Dynamics', Oxford Economic Papers No. 33, 1981.

3. See Note 1.

4. A recent estimate of this effect by the Treasury is that oil has produced a 10-15 per cent increase in the real exchange rate. See Treasury Working Paper No. 22, p. 18.

5. See Note 4.

6. Among the supporters of such a fund have been Dr David Owen (*Financial Times*, 16/1/78); Mr Tony Benn, Energy Secretary, Cmnd. 7143; Mr Jack

Jones (*Financial Times*, 4/12/77); and Mr T. Eggar (*The Times*, 2/11/81).

7. H.C. 720, pp. 56-8.

8. In 'Does Monetarism Fit the UK Facts', by A. Budd, S. Holly, A. Longbottom and D. Smith. Presented at the City University Conference on Monetarism in the UK, September 1981.

9. In the Treasury and Civil Service Committee Report on Monetary Policy, 1979-80, H.C. 720, p. 53.

10. See 'Does Monetarism Fit the UK Facts', p. 56.

11. From a speech by Mr Burns, Chief Economic Adviser to the Treasury, in Washington, 24 September 1981, p. 7.

12. In the third Report from the Treasury and Civil Service Committee, Volume III, H.C. 163-III, p. 92.

13. Ibid., p. 91.

14. Ibid., p. 95.

15. M. Beenstock and A. Longbottom, in *Economic Outlook*, June 1980.

16. See the National Institute *Economic Review*, November 1980, No. 94, p. 19.

17. A more detailed discussion of this may be found in C. Taylor and A. Threadgold (1979), Bank of England Discussion Paper 6, or M. Miller (1982), Warwick Economic Research Paper No. 209.

7 THE GOVERNMENT'S MEDIUM-TERM STRATEGY

Introduction

When North Sea oil was in prospect it was widely expected that self-sufficiency in oil would bring in a period of prosperity for the British economy. In fact the early 1980s have been years of deep recession. There are various explanations for this contrast between expectations and reality, including the slowdown in world economic activity connected with the second wave of OPEC price increases. But on our interpretation of events a major reason for the UK's slide into recession at a time of rapidly rising oil production was the restrictive economic policy adopted by the Conservative government which took office in May 1979. This policy was elaborated in the 'medium-term financial strategy.'

The Historical Background

Much of the postwar period of British economic history has been dominated by the 'stop-go' cycle. Successive governments adopted expansionary fiscal policies in order to increase employment levels: as output rose the balance of payments would move into deficit and, under the Bretton Woods system of relatively fixed exchange rates, the only cure for this deficit was to reverse the fiscal expansion and allow unemployment to rise. In November 1967 an attempt was made to break out of this cycle by devaluing within the Bretton Woods system, and the system itself broke down in the early 1970s with the floating of sterling, the dollar and many other currencies.

Underlying this stop-go cycle, however, in the present government's view, was a more serious long-term problem. Figure 7.1 shows unemployment levels from 1950-79; the cycles of unemployment, rising and falling, are quite evident but there is also a long-term upward trend in unemployment. Each peak of unemployment is higher than the one before. There also appears to have been a rise in the rate of inflation from 1959 onwards (Figure 7.2). Figures 7.3 and 7.4 show that the

Figure 7.1: Total Unemployment in the UK, 1950-1979

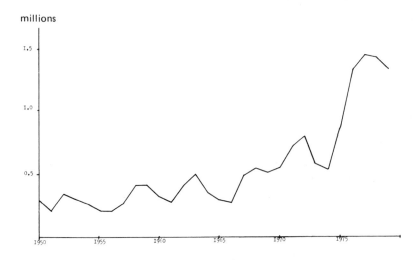

Figure 7.2: The Rate of Increase of the Retail Price Index, 1950-79

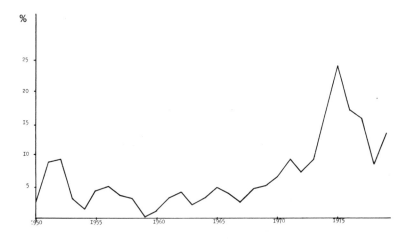

Figure 7.3: Percentage Rate of Growth of Real GDP

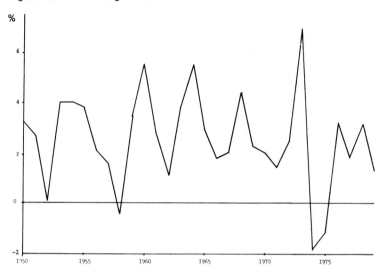

Figure 7.4: Percentage Rate of Growth of Real Manufacturing Production

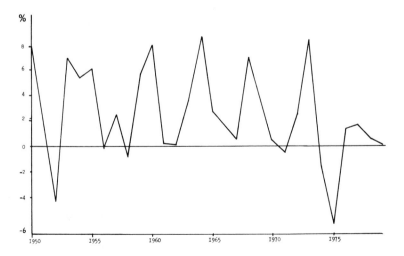

growth rates of GDP and manufacturing production were fairly low in the 1970s.

The picture which the government had formed of the problems facing the economy was one of a serious underlying long-term supply problem.

> The poor performance of the British economy in recent years has not been due to a shortage of demand. We are suffering from a growing series of failures on the supply side of the economy.[1]

The government set itself two objectives. Inflation had to be overcome, and progress had to be made towards restoring incentives and improving the efficiency of industry. It was felt that the latter changes could be achieved by restructuring the tax system and generally reducing the scale of government intervention. The Chancellor made it clear, however, that 'these changes will not themselves be enough unless we also squeeze inflation out of the system'.

The government's interpretation of events was therefore one of a progressively worsening situation. This interpretation is perhaps unduly pessimistic. Much of the deterioration in the trends outlined in Figures 7.1 to 7.4 results from the recent adjustment to the 1973-4 oil crisis. The unemployment figures from 1950-73 show only a slight suggestion of a long-term trend. The peak rate of growth of real GDP over the whole period occurred in the early 1970s, and there was not even a fall in the overall growth rate of manufacturing until 1974. The real problem concerned the rising rate of inflation.

The major issue, however, is not whether the performance of the economy had been worsening, but whether the new government's policies were such as to improve that performance.

The Medium-term Financial Strategy

The medium term financial strategy (MTFS) was introduced in the March 1980 budget. 'The Government's objectives for the medium term are to bring down the rate of inflation and to create conditions for a sustainable growth of output and employment'.[2] The basic policies had been set out in the budget of June of the previous year; the MTFS was merely a development of this budget. Given its two main aims of reducing inflation and improving the supply side of the economy, the government's strategy has proceeded in two fairly distinct

Table 7.1: Recent Targets for the Growth of £M3

Date set	Period set for	Target range (%)	Out turn (%)
Dec. 1976	12 months to April 1977	9-13	7.7
Mar. 1977	12 months to April 1978	9-13	16
Apr. 1978	12 months to April 1979	8-12	10.9
Nov. 1978	12 months to Oct. 1979	8-12	13.3
June 1979	10 months to April 1980	7-11	10.3
Nov. 1979	16 months to Oct. 1980	7-11	17.8
Mar. 1980	14 months to April 1981	7-11	22.2
(Reissued	12 months to April 1982	6-10	—
in Mar.	12 months to April 1983	5-9	—
1981)	12 months to April 1984	4-8	—
Mar. 1982	12 months to April 1983	8-12	
(General	12 months to April 1984	7-11	
monetary	12 months to April 1985	6-10	
targets)			

directions; the control of inflation has been attempted almost entirely by trying to control the money supply, while a number of measures were introduced to increase incentives and general market freedom. Reducing government expenditure was expected to strengthen the market system as well as helping in the reduction of the money supply.

The policy of having target growth rates for the money supply was not, of course, first applied by the present administration. Indeed there has been an unbroken succession of targets since Mr Healey introduced the first set in 1976. Table 7.1 summarises the recent history of monetary targets in the UK. But the previous government treated the monetary target as only a small part of its economic policy; its main instrument for inflation control was an incomes policy which broke down in the winter of 1978-9. After this the present government's primary way of controlling inflation was to adhere to the set target. Since great stress was laid on meeting this target much of the rest of the government's actions had to fit in with it. The monetary targets outlined in the March 1980 budget related specifically to Sterling M3 and were reaffirmed in March 1981. The 1982 budget made a substantial change in monetary targets, the actual target range was raised and, more importantly, the target was no longer defined in terms of a specific monetary aggregate.

The most important and direct influence which the monetary targets have had on other government policies has been through a companion set of targets for the public sector borrowing requirement. These were introduced informally in the 1979 budget when the Chancellor set a PSBR target for that year of £8¼ billion, a reduction of £1 billion over the year before, which would reduce the PSBR from 5½ per cent of GDP to 4½ per cent. In the budget speech subsequent reductions were then promised. The intention to reduce the PSBR progressively was reaffirmed by the Financial Secretary to the Treasury in January 1980: 'We have made it a central objective of our policy to achieve a downward trend in the PSBR as a percentage of GDP in the medium term.'[3]

Table 7.2: PSBR Targets and Out-turns

Date set	79/80	80/81	81/82	82/83	83/84	84/85
June 1979	4½%					
March 1980	—	3¾%	3%	2¼%	1½%	—
March 1981	—	6%	4½%	3¼%	2%	—
March 1982				3½%	2¾%	2%

Source: Financial Statement and Budget Report, 1979-80, 1980-1, 1981-2, 1982-3.

The detailed plan for this gradual reduction in the PSBR was first outlined in the 1980 budget and subsequently revised in the 1981 and 1982 budgets. Table 7.2 shows details of the PSBR targets and the actual out-turns. The 1980 PSBR plan quickly proved far too ambitious, as the drastically revised 1981 plan illustrates. The 1982 plan was only slightly higher than the 1981 plan. Partly in order to meet these PSBR targets an overall reduction in public expenditure was also planned. In the 1979 budget a planned reduction in spending of £4 billion at current prices was announced. The two subsequent budgets saw more detailed plans for the progressive reduction in government expenditure. Table 7.3 presents the expenditure plans outlined in the 1980, 1981 and 1982 budgets. Table 7.4 shows the development of government expenditure from 1970 to 1980.

It is interesting that the 1980 budget envisaged expenditure declining from 1980 onwards, while the 1981 budget estimates do not peak until two years later. The 1982 budget planned very little decline in expenditure at all: there is a slight peak in 1982/3 but the pattern of decline seems to have disappeared.

Table 6.2 of Chapter 6, which outlines the changes which have occurred in the fiscal stance between 1974/5 and 1981/2, gives an

Table 7.3: The Medium-term Financial Strategy Expenditure Plans

	79/80	80/81	81/82	82/83	83/84	84/85
1980 survey prices £ billion						
March 1980	79.5	78.5	76.5	75.5	75.5	—
March 1981	77.9	79.0	79.5	78.0	76.5	—
March 1982	—	78.7	79.5	81.2	80.0	80.0

Source: Financial Statement and Budget Report, 1980-81, 1981-2, 1982-3.

immediate indication of how deflationary government policy has been. On a standardised basis of 2.5 per cent unemployment the PSBR in 1978/9 would have been 4.4 per cent of GDP. The change in fiscal stance in that year, the last of the Labour government, was actually positive; in other words government policy was expansionary. Since that time the present government has introduced a succession of severe budgets. The change in fiscal stance from 1979/80 became increasingly negative. Over the whole period 1978/9 to 1981/2, if the level of output had remained high enough to keep a constant 2.5 per cent level of unemployment, the PSBR would have changed from +4.4 per cent of GDP to −2.4 per cent. This illustrates the restrictive nature of government policies.

Table 7.5 shows the breakdown of public expenditure into its constituent parts. It is apparent that the government succeeded in cutting large areas of its expenditure. As proportions of total current expenditure, wages and salaries have fallen, subsidies to industry have fallen, as have foreign grants, and there has been a dramatic fall in public capital expenditure between 1979/80 and 1981/2. But at the same time as these areas have contracted, there has been a substantial rise in transfers to the private sector. This covers most of the unemployment-related payments, such as social security. While the non-recession-related sides of government expenditure have been contracting, the unemployment-linked payments have risen dramatically.

Even if the personal sector alone is considered the MTFS has been restrictive. This is illustrated in Figure 7.5 which shows the movements in the Retail Price Index and the Tax and Price Index. This second index was introduced by the government to show the position of individuals when changes in direct taxation were taken into account in the calculation of inflation rates. When in the initial budget VAT was raised and direct taxes were reduced it was argued that the RPI

Table 7.4: Government Expenditure, 1970-80

£billion 1975 prices	70	71	72	73	74	75	76	77	78	79	80
Total expenditure	40.9	41.5	42.3	45.4	49.8	51.6	51.3	48.3	50.4	52.5	54.3
Expenditure on goods and services	24.6	25.0	25.6	27.3	27.2	28.0	28.0	27.0	27.0	27.1	27.3
Non-dwelling public investment	6.9	6.8	6.3	7.2	7.1	6.9	6.7	5.8	5.3	5.1	5.0

Source: Economic Trends Annual Supplement, 1981.

Figure 7.5: The Annual Rate of Change of the RPI and the Tax and Price Index

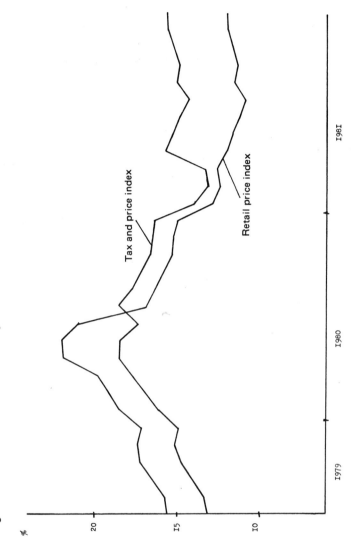

Table 7.5: Breakdown of Public Expenditure Changes

	76/77	77/78	78/79	79/80	80/81	81/82 (Estimated outturn)	82/83 (Plan)
	Categories of current public expenditure as % of total current expenditure						
Wages & salaries	39.5	38.0	37.0	36.6	37.4	36.4	34.8
Goods & services	20.3	20.5	20.4	20.8	20.7	20.8	21.6
Subsidies	8.5	7.1	7.0	7.3	7.4	6.4	5.7
Transfers to private sector	29.9	31.9	32.7	32.6	32.8	35.0	36.2
Grants abroad	1.8	2.5	2.8	2.6	1.6	1.2	1.6
	Total public capital expenditure as % of total public expenditure						
	16.8	13.0	13.4	16.4	12.8	10.9	10.2

Source: Government's Expenditure Plans 1982-3 to 1984-5, Cmnd. 84 94-I.

would not correctly measure inflation over the period, as this index captures only the change in indirect taxation. For the first year the new Tax and Price Index was substantially lower than the RPI because of the reduction in direct taxation. After the March 1980 budget, however, the relationship between the two changed and the TPI rose faster than the RPI. There was a further sudden jump in the TPI after the March 1981 budget failed to raise a range of tax allowances.

It is difficult to assess what the government's expectations were for the medium-term financial strategy. The only quantitative forecasts which it has published are very short-term forecasts generated by the Treasury model which is not a monetarist model. The following statement by the Chancellor in his first budget suggests that these forecasts were not what the government expected to happen.

The conventional forecasting arithmetic, which, in accordance with custom and statute, I am publishing in the financial statement, does suggest that the economy will show no growth in the period immediately ahead.

But this prospect, insofar as it can be viewed as a reliable prediction — which itself is open to doubt — cannot be taken to mean that the budget is, in the traditional language of neo-Keynesian economists, perversely contractionary.[4]

The Chancellor finished his 1979 budget speech on a note of high optimism.

> The Budget is designed to give the British people a greater opportunity than they have had for years to win a higher standard of living . . . I dare to believe that they will respond to the opportunity that I have offered them today.[5]

The government's hopes for its economic strategy were high if not for the current year then for the period 2-3 years ahead. This point is illustrated by the following passage from the 1981 Budget Report, commenting on the performance of the economy over the previous year: 'The recession has been worse than was expected a year ago, and in particular manufacturing output and employment have fallen sharply.'[6] This suggestion, that the government had not expected the large fall in output and employment, is further illustrated by the revised PSBR targets in the 1981 budget. The projections in the 1980 budget were founded on the assumption of an average rate of growth of 1 per cent from 1981 onwards, but this was qualified by the following statement: 'The economy should be capable of growing faster than this, but for the purpose of these projections it is prudent to assume a low growth rate.'[7]

There have been four budgets under the influence of monetarism: the June 1979 budget and the three succeeding March budgets. The first two set out the initial MTFS; this aimed at a sharp reduction in the rate of growth of sterling M3, a fairly rapid decline in government expenditure and a sharp fall in public borrowing. The subsequent two budgets have contained some revision of these policies. In 1980 government expenditure was expected to peak in 1979/80 and then decline substantially; by 1982 the expenditure plan for 1984/5 was actually higher than the peak figure in the 1980 plan. Similarly the monetary targets set out in November 1979 and March 1980 were exceeded by such a large amount that the 1982 targets were raised substantially and made more vague by not targeting a specific definition of money. The only area of the MTFS which has remained intact is the PSBR targets. These have been raised slightly in each successive year, but the overall pattern of decline still remains. The supply side objectives of the MFTS in 1980 have been almost totally abandoned. The non-indexation of tax allowances in the 1981 budget reversed much of the effect of the earlier tax concessions, which were originally intended to be only a first step.

The Development of the Economy under the MTFS

Figures 7.6 to 7.10 show that since the second quarter of 1979 the UK economy has moved into what can only be regarded as a major recession. Table 7.6 puts this recession in perspective by comparing it with the cyclical behaviour of the economy since 1950. It is immediately clear that the scale and depth of this recession is in no way comparable to what might be termed normal economic fluctuations. Throughout all the cycles from 1950 to 1970 gross domestic product actually rose in absolute terms even during the periods of contraction. The overall growth trend of the economy was strong enough over this period to dominate the 'stop-go' cycle. The 1973 oil crisis precipitated the first recession which led to an actual fall in GDP but even this decline was only half the decline which has occurred in the current recession so far. The rise in unemployment is also more than twice that in previous recessions.

Figure 7.6 shows that the rate of growth of GDP slowed down after the second quarter of 1979 and became heavily negative in the second quarter of 1980. Figure 7.10 shows an even more dramatic decline in manufacturing output while Figure 7.7 shows the corresponding sharp rise in unemployment over the period. Figures 7.8 and 7.9 show an interesting pattern between the money supply and the rate of inflation. The RPI rose rapidly after the 1979 budget largely as a result of the increase in VAT and then the subsequent increase in oil prices. The money supply began to rise dramatically in the second quarter of 1980, despite the government's attempts to meet its monetary targets. This lag in the response of the money supply to the price level is exactly what would be expected from our arguments put forward in the last chapter. Despite attempts to control the money supply the quantity of money increased rapidly in order to accommodate the recent price rises.

The unique severity of the current recession is quite apparent, but this does not necessarily mean that the medium-term financial strategy is responsible for it. There are other possible explanations, such as the massive increase in world oil prices. It is not possible to assess the relative importance of the various possible causes on a purely theoretical

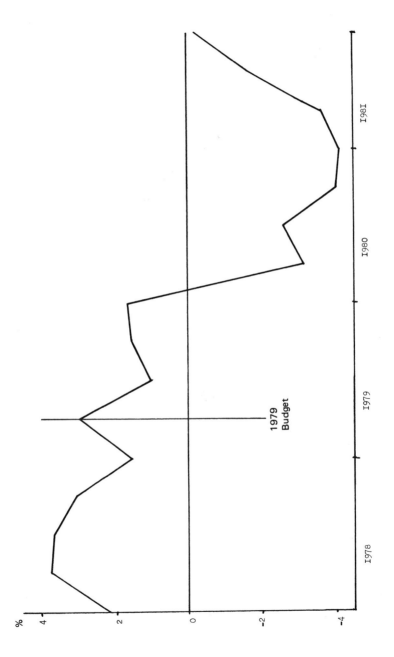

Figure 7.6: Percentage Rate of Growth of Real GDP, 1978-81

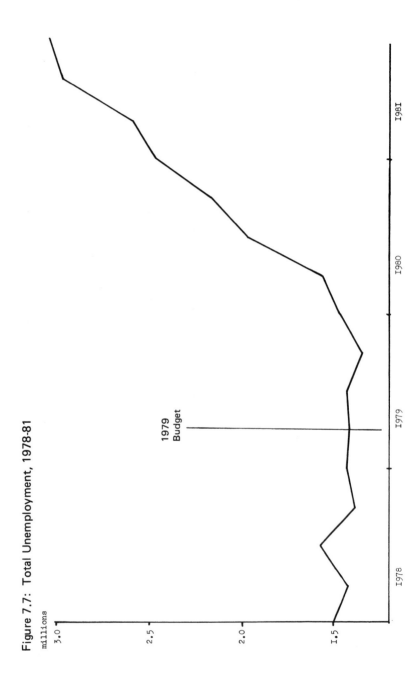

Figure 7.7: Total Unemployment, 1978-81

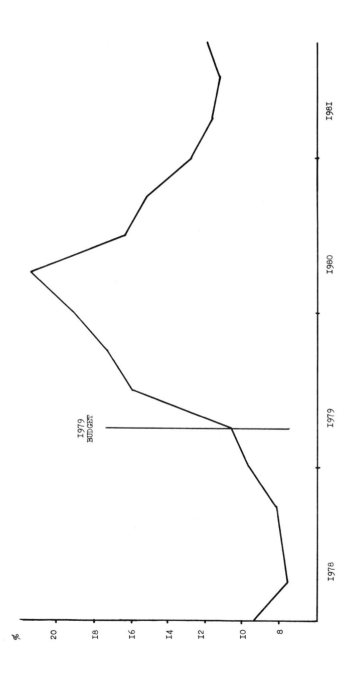

Figure 7.8: Percentage Annual Rate of Increase in the RPI

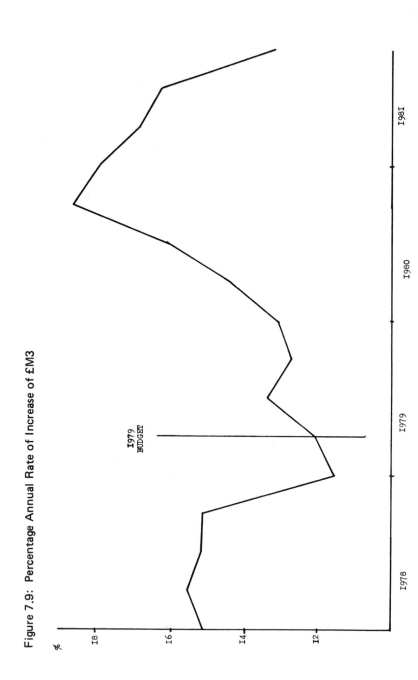

Figure 7.9: Percentage Annual Rate of Increase of £M3

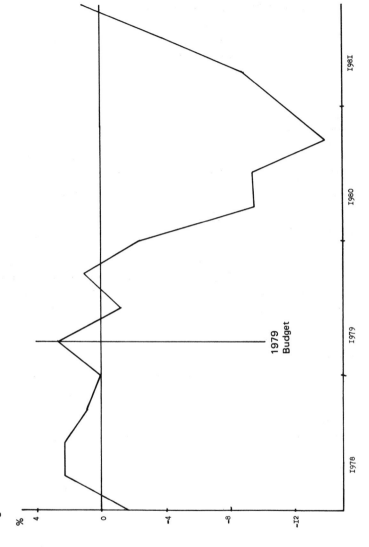

Figure 7.10: Annual Growth Rate of Manufacturing, 1978-81

Table 7.6: Economic Recessions, 1951-81

Recession		GDP % change peak to trough	Industrial production % change peak to trough	Unemployment %		
Peak	Trough			Peak	Trough	Change
1951I	1952III	—	—	1.2	1.7	+0.5
1955IV	1958III	3.1	−1.2	0.9	2.0	+1.1
1960III	1963I	3.3	−0.4	1.5	2.4	+0.9
1965I	1967III	5.0	2.6	1.3	2.3	+1.0
1969I	1972I	4.6	−2.4	2.3	3.9	+1.6
1973II	1975III	−3.8	−10.1	2.7	4.1	+1.4
1979II	1981II	−7.5	−14.1	5.4	10.4	+5.0

Source: National Institute Economic Review, November 1981.

basis. Instead an empirical model must be used to assess the possible course of events if policy had been broadly 'neutral'. Both the London Business School and the National Institute carried out such an exercise for the Bank of England's Academic Panel in July 1981.[8] Their results are summarised in Table 7.7.

Each group made slightly different assumptions but their overall results are very similar. If responsibility is assigned to the government for the exchange rate as well as its direct fiscal and monetary action, then the National Institute finds that the government was responsible for 73 per cent of the overall fall in output while the London Business School assesses the figure at 67 per cent. This exercise has recently been updated by the National Institute to include the large fall in output which occurred in 1981. The results of this exercise were almost identical to the original in the sense that the largest contribution to the shortfall in GDP was found to be direct government action. This result has recently been confirmed by Henry and Wren-Lewis.[9] They estimated a reduced form output equation which was a function of lagged output, lagged government fiscal policy and the lagged real exchange rate. This equation was estimated over the period 1972 to 1979, and it was found to explain almost all the fall in output which occurred in 1980 and 1981. This means that it is possible to explain almost the whole of the current recession simply by considering the effects of government fiscal policy and of the high exchange rate. It is, of course, always necessary to stress the large degree of uncertainty which must be attached to these empirical results. Nonetheless, the

Table 7.7: Explanations of the Shortfall in Output, 1978-80 (%)

	NIESR	LBS
Non-recessionary trend in output	3.6	3.2
Actual change in output	−0.5	−0.8
Shortfall	4.1	4.0
of which:		
policy effects, i.e. public spending not growing in line with trend output; changes in (indexed) tax rates; changes in the level of 'real' interest rates	1.7	1.6
exchange rate being other than would maintain constant competitiveness	1.3	1.1
increase in real oil price and world prices and their effects on world activity	0.8	1.3
real wages in the United Kingdom growing faster than trend output	–	0.1
Total identified	3.8	4.1
Unexplained residual	0.3	−0.1

Source: *Factors Underlying the Recent Recession*, page 7. Bank of England, Papers presented to the Panel of Academic Consultants, no. 15.

evidence seems strongly to indicate that the prime responsibility for the current recession lies with government policy.

Even in his inaugural budget speech, in 1979, the Chancellor recognised that his policies would lead to a short-term fall in the rate of growth of the economy: 'Any decline in economic activity which might, on a narrow view, be attributed to this Budget will be essentially the consequence of the economic situation which has made such measures inevitable.'[10]

It is, of course, also true that the Chancellor expected this decline to be only temporary: 'I have stressed the urgent need for new policies to reverse the decline of the British economy'.[11] The final verdict on the government's economic strategy must, however, wait on the development of the economy over the next three to four years.

In order to achieve a better assessment of the long-term effect of the MTFS the next section will consider the likely development of the economy up to 1986.

The Likely Development of the Economy under the MTFS

In order to make an overall assessment of the likely success of the MTFS we must have a forecast of the likely developments of the economy under these policies. The National Institute of Economic and Social Research regularly produces forecasts which are carried out on precisely this 'unchanged policy' assumption. That is to say the Institute outlines the likely development of the economy if current government policies were maintained over the forecast period. In many ways this forecast is more appropriate in assessing the success of a set of economic policies than actual foresight, as it is always possible that future events will be affected by factors which are entirely unknown. For example, a major boom or collapse in the level of world trade might easily have repercussions on the UK economy such as to appear either to validate or destroy the MTFS. But if such an occurrence was unforeseen by government policymakers it would hardly be credible to ascribe this effect to the policy itself. A forecast made at the same time as the policy has the advantage, therefore, of having available to it only the same set of information as the policymaker. The disadvantage of using a forecast is, of course, that forecasts actually differ and are themselves based on an underlying framework which is either hostile or sympathetic to the MTFS. The National Institute's forecast has been chosen for presentation here for a number of reasons; firstly, it is independent of any direct government influence; secondly, the Institute has been producing regular published economic forecasts for a very long period of time and has consequently built up considerable expertise and experience in the field; finally, the medium-term forecasts, which are regularly published, cover a five-year period which is particularly useful in assessing the longer-term performance of a set of policies.

The National Institute's November 1981 forecast has been chosen as the base case for assessing the MTFS strategy. This forecast covered a five-year period up to the first quarter of 1987. It was first published in the National Institute's Economic Review in November 1981 and a

Table 7.8: Simulation of the National Institute's Model on the MFTS Policy Assumptions

Year	Real GDP % change	Unemployment million 4th quarter	Retail prices % change	PSBR £bn	% GDP	£M3 % change
1979	2,1	1.3	13.4	12.6	7.4	12.7
1980	−2.8	1.9	17.9	12.3	6.4	18.6
1981	−3.1	2.8	12.0	11.0	5.3	11.5
1982	0.6	3.0	10.7	11.0	4.8	9.5
1983	1.4	3.1	8.3	10.0	4.0	9.0
1984	2,1	3.2	8.0	7.0	2.6	8.5
1985	1.5	3.2	7.9	4.0	1.3	7.0
1986	0.9	3.4	8.0	2.0	.6	6.5

Table 7.9: Further Details of the Simulation

Year	GDP Index 1975 = 100	Stock building £M 1975	Manufacturing index 1975 = 100	Current balance £ 1975
1979	110.3	1,504	104.6	−.6
1980	107.2	−1,978	94.8	1.6
1981	103.9	−1,849	88.9	1.8
1982	104.5	204	89.7	1.9
1983	105.9	480	90.3	3.0
1984	108.2	776	91.6	3.2
1985	109.8	599	92.8	3.9
1986	110.8	531	93.9	4.5

more detailed account of its assumptions and background reasoning may be found there. The policy simulations presented in Chapter 8 will use this forecast as a base.

Table 7.8 presents a summary of the National Institute forecast results for the period 1979-86. At first glance this seems to imply a fair degree of success for the MTFS. Its intermediate targets for the money supply and the PSBR are nearly met while the primary economic objective of reducing inflation is fulfilled. The figures for the growth in real GDP and unemployment present a rather different picture, however. Unemployment continues to rise over the whole period, while the projected growth rate for GDP is positive but very low indeed. Table 7.9 presents some further figures from the forecast which help to assess the overall development of the economy.

Between 1970 and 1980 the average rate of growth of the GDP was 1.4 per cent; if this average had continued beyond 1979 the GDP index would have stood at 121.5 per cent in 1986. The forecast GDP index for 1986 in fact stands at a figure which is almost identical to that of 1979; even though GDP has been rising from 1981 onwards this rise has only just been sufficient to make up for the fall in GDP from 1979-81. In 1986 GDP is still 10 per cent below the level it would have reached if the trend rate of growth of the 1970s had been maintained.

The fall in output in 1980 and 1981 primarily took the form of a large decline in stockbuilding. It is not possible, however, for stocks to decline indefinitely. In the normal cycle of booms and slumps the end to de-stocking often marks the end of the recession. This is not the case, however, in the National Institute forecast. It cannot really be said that the recession ends at any time in the forecast period. What happens is that the recession is particularly bad in the early years, while stocks adjust, but that a low trend growth reasserts itself in 1982-3. This entails a low but steady rate of growth which is not sufficient to reduce unemployment levels. The high growth rate which would be typical of the move into a boom period simply does not occur. This pattern is illustrated clearly by the manufacturing output index which shows a large drop between 1979-81 and then a very slow but steady rise.

Table 7.10 gives the value of oil and gas production and the level of government tax revenues generated by the sector. These figures affect the forecast in three main ways: through a direct contribution to GDP, by direct increases in taxation, and through the balance of payments. The value of oil production expressed as a percentage of GDP rises from 4.5 per cent in 1979 to 9.9 per cent in 1986. The overall level of GDP in 1986 is almost identical to that of 1979. The direct impact of the North Sea on GDP is therefore to prevent a fairly substantial absolute decline in GDP occurring over the period. Secondly, the rise in tax revenues from the North Sea between 1979 and 1986 is very large and can be directly linked to the fall in the PSBR. In nominal terms the PSBR falls by £10 billion while North Sea taxes rise by £17 billion. The earlier discussion of the MTFS expenditure plans made it clear that the government has failed to achieve its planned expenditure cuts. The attainment of its PSBR targets must therefore require a rise in total taxation. The large rise in North Sea

Table 7.10: The Importance of Oil and Gas in the Simulation

	Government tax revenues		Total value of oil and gas production	
	£ Bn 1975	£Bn current	£Bn 1975	% GDP
1979	1.2	1.5	4.7	4.5
1980	1.4	3.3	6.4	6.3
1981	1.9	3.8	7.5	7.6
1982	3.0	8.0	8.6	8.7
1983	4.2	11.2	9.2	9.2
1984	5.0	13.9	9.6	9.4
1985	5.4	15.6	10.2	9.8
1986	5.8	18.6	10.4	9.9

revenues is therefore directed into reducing the PSBR to a level roughly in line with the MTFS strategy. Thirdly, the balance of payments effects of oil are quite startling. A small deficit in the current balance in 1979 is transformed by 1986 into a large surplus. The rise in the current balance mirrors the rise in the total value of oil production.

A surplus on the current account of the balance of payments must be balanced by the net acquisition of a similar amount of foreign assets, either by the government or by the private sector. A current account surplus is equivalent to investing overseas. The large oil induced surplus shown in the forecast therefore implies that present policies will result in the large scale investment of oil revenues overseas. Oil revenues are therefore being used in two main ways by the present government. Firstly, an attempt to reduce the PSBR as a priority leads to most of the tax revenues being directed towards this end; secondly, because the restrictive fiscal policies have put the economy into a serious depression, a current account surplus has developed which implies the investment of a large part of the oil revenues overseas.

It is not necessarily wrong to invest overseas, but current policies are not consciously directed towards this end. The foreign investment has results both from the oil and from the recession itself, which has reduced the demand for imports generally. If the economy were to be expanded the oil effect would remain in the balance of payments but would be offset by higher imports and the current balance could easily move into deficit. A case can be made for investing overseas, but this should be financed by a surplus on current account which is earned at a high level of activity, not one which arises only from a continuing recession.

How then should the medium term financial strategy be assessed? The answer to this must be in two parts: firstly, the shorter term aim of controlling the money supply and the PSBR, and secondly, the ultimate objective of reducing inflation and improving the supply side of the economy. As mentioned earlier the forecast seems to give some support to the idea that the shorter term objectives of the MTFS will be achieved. This is not altogether correct, however. The PSBR does fall much in line with government plans, but this is not due to a reduction in government spending but rather to an increase in taxes from the oil sector. Total oil taxes are forecast to increase by nearly £15 billion a year between 1981 and 1986 in nominal terms. This means that if all other taxation and expenditure remain unchanged but there was no rise in oil taxation the PSBR in 1986 would stand at £17 billion. The fall in the PSBR is almost exactly in line with the rise in oil taxation. While the PSBR target is met it does not indicate an underlying reduction in the scale of government activity. The forecast suggests that inflation and the money supply will fall fairly constantly over the period. However, it is not a case of success in controlling the money supply causing a reduction in inflation. What is actually happening in the forecast is that the high unemployment levels are causing relatively low wage settlements which in turn are feeding through to a reduction in inflation. The money supply moves to accommodate a demand for money which is predominantly determined by the inflation rate. As inflation falls over time the rate of growth of the money supply also falls. The money supply is not moderating inflation, this is being done by unemployment and the recession. Further, even at the end of the period, inflation is still continuing at some 8 per cent a year. The government has argued that the long-term effects of the MTFS would be a revitalisation of British industry as the supply side of the economy responded to the improved opportunities provided by the free market. In the Institute forecast there is no hint of such a response. Output remains depressed well below its trend levels and the manufacturing sector in particular suffers a serious loss in production. Above all the continued rise in unemployment is not a symptom of a major improvement in the supply side of the economy. Recently there has been a tendency to point to productivity indices as a sign of the expected supply side improvement. These indices have shown a fairly large rise during 1981. However, a productivity index simply measures output per person employed. In time of recession when both output and employment are falling, the movement of the index will depend on whether output falls faster than employment. In fact a severe recession would be expected to raise the normal productivity index because those firms which are driven out of business will on average have low productivity and profitability. When the recession finally ends and demand returns to normal such low profitability enterprises are likely to be restarted, bringing the productivity index back down. A rise in the productivity index does not imply an increase in potential output since a return to full employment would almost certainly be accompanied by a fall (or at least a decline in the trend rate of growth) of the index.

Conclusion

The medium-term financial strategy was effectively introduced in June 1979 against a background of economic events which were interpreted as showing the symptoms of a long-term decline in the general performance of the UK economy. It was felt that inflation was increasingly beyond control, that unemployment had been steadily rising and that the supply side of the economy was suffering from serious distortions. It was recognised that in its immediate effects the MTFS was likely to reduce the growth rate of the economy, but it was expected that over a longer period inflation would be brought under control and the productive sectors of the economy would be brought back to full strength.

Since the introduction of the MTFS the economy has been plunged into the worst recession since the war. Unemployment has risen dramatically and output has fallen in absolute terms. There is little sign that any substantial economic recovery is likely to occur without a significant change in government economic policies. On the basis of current and likely developments the MTFS must be judged to have failed to achieve its objectives of revitalising the British economy. Indeed, it bears a major part of the responsibility for the current economic weakness. Even its more limited aims of reducing inflation have been achieved not through control of the money supply but rather through widespread and continuing unemployment.

In reply to criticism the government has repeatedly said that 'there is no alternative'. The next chapter will examine a number of suggestions for alternative strategies.

Notes

1. Sir Geoffrey Howe in his 1979 Budget speech. Hansard No. 1140, column 240.

2. Financial Statement and Budget Report, 1980-81, p. 16.

3. Speech to the *Financial Times*, 1980 Euromarkets Conference, January 1980.

4. Hansard, No. 1140, 11-14 June 1979, column 243.

5. Hansard, No. 1140, 11-14 June 1979, column 263.

6. Financial Statement and Budget Report, 1981-2, p. 15.

7. Financial Statement and Budget Report, 1980-1, p. 18.

8. *Factors Underlying the Recent Recession*, Bank of England Paper, no. 15.

9. S.G.B. Henry and S. Wren-Lewis, presented at a conference in Paris, 1982.

10. Hansard, No. 1140, 11-14 June 1979, column 244.

11. Ibid., column 262.

8 ALTERNATIVE POLICIES*

In this chapter we consider what economic policies would permit a return to faster economic growth supported by the wealth of the North Sea.

As the recession developed in 1980 various groups of economists proposed alternative policies to promote expansion without exacerbating inflation. The Clare group recommended a cut in the National Insurance surcharge on employers, a cut in VAT, restoration of income tax allowances, extra investment in the nationalised industries, lower prices for public sector goods and services, and increased current government spending. They estimated that the total value of this package would be of the order of £5 bn a year, although the eventual effect on the PSBR would be only £3 bn. No details were given of the likely effects on output or employment, although a general statement was made that 'A government that wants recovery must be willing to run an "inflation-adjusted" deficit in a recession and allow the PSBR to rise as necessary'.

The TUC, in its 1981 and 1982 Economic Reviews, also set out an alternative to the MTFS. Its recommendations were broadly similar to those of the Clare group, an increase in public investment both in private and publicly owned industry, a general increase in government expenditure on education, health, social benefits, pensions and local authority spending, and full indexation of income tax allowances with drastic changes to the National Insurance scheme. In total the TUC package outlined in 1981 represented an injection of £6.2 bn into the economy in 1981-2. The detailed effects were, once again, not spelt out, although the TUC argued that 'This is the only way for the economy to break out of the current straitjacket in which the government has imprisoned it, and to turn the economy around from its present downward spiral on to a course of economic growth, higher investment, higher employment and rising living standards.' The 1982

---

* These projections and simulations were obtained with the help of the NIESR model (vintage 1981). The Institute itself has no responsibility for the way its model has been used in this book, nor for any of the results we have obtained.

179

economic review suggested a slightly higher reflation of £8.3 bn, although part of the increase was simply due to inflation.

In early December 1981 Professor Sir Bryan Hopkin, with Professor Marcus Miller and Professor Brian Reddaway, put forward a reflationary package which was basically similar to the Clare group's suggestions. The National Insurance surcharge was to be abolished, both the exchange rate and interest rates were to be reduced, VAT was to be reduced, a general increase in the capital expenditure of the nationalised industries and a reduction in nationalised industry prices were to be effected. The total value of this package was assessed at £6.8 bn and the ultimate increase in the PSBR at only £1.3 bn. This proposed alternative differed from the first two in that it included a fairly detailed assessment of the likely effect of the proposals after one year. The expected effect on GDP was an increase of 3 per cent which would lead to extra employment of 480,000 and this would reduce the recorded unemployment figures by 320,000. The reduction in VAT would lead to 2.4 per cent less inflation, while they expected only negligible effects on wage settlements. The current account of the balance of payments was actually expected to improve due to the devaluation of sterling, by £0.4 bn. One important fact which emerges immediately from these figures is that such a reflation is not a quick cure for recession. Given the predicted unemployment level for 1983 of 3.1 million (see Chapter 7) a reduction of 300,000 is really only preventing the rise in unemployment. The longer-term effects of the policy are presented in only general terms. 'We would emphasize the fact that these are only the first year results of our alternative strategy. The measures listed would produce further benefits in later years and, much more important, it should be possible to introduce further measures in succeeding years.'

These alternative strategies have a great deal in common both in terms of the size of the reflationary stimulus and the way in which it should be implemented. The next section will consider the medium-term effects of this kind of stimulus in more detail.

The Medium-term Effects of Reflation

This section will analyse the effects of various economic policy changes over a medium-term time horizon by using the macroeconomic model of the National Institute. The simulation presented in the last chapter for the likely development of the economy on an unchanged policy basis will form the base case for the simulations presented here. That is to say we will be investigating the effects of various policy changes given the basic development of the economy outlined in the last chapter. The policy changes will begin in the first quarter of 1981 so before this time the development of the economy is identical to the base case. The choice of the starting date for the simulation is not easy. An earlier starting date might have been chosen, in particular we might have altered government policy from the start of this government's term of office in mid-1979. The argument against this is that the development of the economy between 1979 and 1981 would then have been so different that the recession would hardly have appeared at all. But our major objective here is not to show that things might have been different but to show how a major recession may be dealt with. By using a starting date of 1981 we have allowed the recession to take full hold of the economy before implementing any policy changes. Alternatively, we could have adopted a later one, preferably a starting date which would be later than the publication of this book. By doing this the result would have been more relevant. The disadvantage would be that if the policy changes did not occur until 1983 or 1984 the forecasting period would have to be extended to 1988 or 1989 to cover the medium term; this would reduce the reliability of all the simulations. Taking 1981 for a base year means that the first year is one for which figures are available.

The suggestions outlined earlier all amount, broadly, to a reflationary package of the order of £5-£6 bn with the major channel being the abolition of employers' National Insurance contributions. The following simulations will investigate the medium-term effect of a reflation of this order through a number of different channels. A single change in a policy instrument will be made, of the order of £5 bn in each simulation, in order to assess the effectiveness of different types of policy changes. The abolition of National Insurance employers' contributions will form the basis of Simulation 1. In fact, the total value of these contributions is something under £4 bn, but for comparison with later simulations the reflation will be carried out at a level of £5 bn. The excess above the actual level of contributions may be seen simply as an employment subsidy.

Simulation 1

The impact on the real economy of reducing employers' National Insurance contributions by £5 bn is very small. The output and employment effects can really be regarded as negligible, the only sizeable

Simulation 1: A £5 billion Reflation Through a Reduction in Employers' National Insurance Contributions from 1981 Onwards

Year	Real GDP % change	Unemployment millions 4th quarter	Retail prices % change	PSBR £Bn	% GDP	£M3 % change
1979	2.1	1.3	13.4	12.6	7.4	12.7
1980	−2.8	1.9	17.9	12.3	6.4	18.6
1981	−2.8	2.7	10.4	16.8	8.3	10.9
1982	1.1	2.8	8.6	13.5	6.1	6.8
1983	1.3	2.9	6.6	14.1	5.9	8.5
1984	1.9	3.1	6.6	13.6	5.3	6.6
1985	1.3	3.2	6.7	11.5	4.1	5.5
1986	0.8	3.3	7.2	10.9	3.7	4.6

Year	GDP index 1975 = 100	Stock building 1975 £M	Manufacturing index 1975 = 100	Current balance £Bn 1975
1979	110.3	1504	104.6	−.6
1980	107.2	−1978	94.8	1.6
1981	104.2	−1841	89.1	1.8
1982	105.3	312	90.1	1.5
1983	106.7	626	90.8	2.7
1984	108.7	1005	92.0	3.0
1985	110.2	776	93.1	4.0
1986	111.0	577	94.1	4.7

effects being on the current balance and the rate of inflation. This instrument works primarily on prices, with only very small output effects. In the initial year the PSBR increases by £6.8 bn.

The orthodox prescription for raising employment is to increase government expenditure or cut taxes. The following two simulations will outline the effects of increasing government consumption and government investment.

Simulation 2

A simulated expansion in government consumption of £5 bn from 1981 onwards has a large impact in 1981 itself. The rate of decline of GDP falls from 3.1 per cent to only 1.7 per cent while unemployment is 400,000 less. The surplus on the balance of payments is reduced. After

Simulation 2: An Increase in Government Expenditure on Current Goods and Services of £5 billion

Year	Real GDP % change	Unemploy- ment millions 4th quarter	Retail prices % change	PSBR £bn	PSBR %GDP	£M3 % change
1979	2.1	1.3	13.4	12.6	7.4	12.7
1980	−2.8	1.9	17.9	12.3	6.4	18.6
1981	−1.7	2.4	12.1	15.4	7.3	13.6
1982	0.6	2.6	10.9	13.6	5.8	11.5
1983	1.5	2.8	8.5	13.4	5.2	11.3
1984	2.0	2.9	8.2	11.7	4.2	10.7
1985	1.4	2.9	8.0	9.1	3.0	9.1
1986	0.9	3.1	8.1	7.6	2.3	8.7

Year	GDP index 1975 = 100	Stockbuilding £M 1975	Manufacturing index 1975 = 100	Current balance £Bn 1975
1979	110.3	1504	104.6	−.6
1980	107.2	−1978	94.8	1.6
1981	105.3	−1811	89.4	0.9
1982	106.0	218	90.4	0.9
1983	107.5	547	91.1	1.9
1984	109.7	958	92.3	2.1
1985	111.2	761	93.4	2.8
1986	112.2	587	94.5	2.4

the initial impact of this fiscal change the economy develops in much the same way as under the base case policies. That is, unemployment rises steadily although at a uniformly lower level, and output rises at much the same rate although from a slightly higher base. The PSBR rises by £4.4 bn in 1981 but as a percentage of GDP it still does not reach its 1979 level, and it falls quite substantially both in nominal terms and as a percentage of GDP over the period. There is only a very small increase in the rate of inflation from 1981 onwards.

It was noted earlier that most of the decline in the volume of government expenditure which occurred during the 1970s came in the investment sector. Any general reflationary package should in our view entail a restoration of public investment levels. The next simulation will investigate the effect on the economy of an increase in public investment of £5 bn.

Simulation 3

An increase in government investment would affect the economy in much the same way as an increase in government consumption. In 1981 unemployment is 200,000 higher under the investment stimulus although by 1984 the discrepancy has completely disappeared. The PSBR is slightly lower throughout the period in the investment simulation. Such small differences in the simulations are as likely to reflect the properties of the model as they are to reflect the properties of the real world. The overall effect of the stimulus is more important and in this respect there is little significant difference between the two.

Simulation 3: A £5 billion Increase in Public Investment

Year	Real GDP % change	Unemploy- ment millions 4th quarter	Retail prices % change	PSBR £bn	%GDP	£M3 % change
1979	2.1	1.3	13.4	12.6	7.4	12.7
1980	−2.8	1.9	17.9	12.3	6.4	18.6
1981	−2.1	2.6	12.0	14.5	6.9	13.8
1982	0.6	2.7	10.9	12.4	5.4	11.5
1983	1.3	2.9	8.4	12.1	4.8	11.1
1984	2.1	2.9	8.1	10.3	3.7	10.6
1985	1.4	3.0	8.0	7.4	2.4	9.1
1986	0.9	3.1	8.0	5.7	1.7	8.6

Year	GDP index	Stockbuilding £M 1975	Manufacturing index 1975 = 100	Current balance
1979	110.3	1504	104.6	−.6
1980	107.2	−1978	94.8	1.6
1981	104.9	−1695	90.0	1.0
1982	105.5	311	91.0	1.0
1983	107.0	526	91.7	2.1
1984	109.2	790	93.0	2.2
1985	110.7	599	94.1	2.9
1986	111.7	528	95.2	3.5

The next two simulations portray the effect of a £5 bn reduction in taxation, the first a reduction in indirect tax (VAT) and the second a reduction in direct taxation.

Simulation 4

A £5 bn cut in indirect taxes has a much smaller effect on output and unemployment than an equal increase in government expenditure. Its main impact is on the level of inflation, which is reduced considerably. Lower UK prices relative to the rest of the world give rise to some substantial balance of payments effects: the current balance is reduced between 1981-3, it is unchanged in 1984 and it is subsequently higher: the J-curve effect.

Simulation 4: A Reduction in Indirect Taxation of £5 billion

Year	Real GDP % change	Unemploy- ment millions 4th quarter	Retail prices % change	PSBR £bn	PSBR %GDP	£M3 % change
1979	2.1	1.3	13.4	12.6	7.5	12.7
1980	−2.8	1.9	17.9	12.3	6.4	18.6
1981	−2.4	2.7	9.3	15.4	7.6	11.4
1982	0.7	2.9	8.7	14.9	6.8	8.4
1983	1.3	3.0	6.5	14.6	6.2	7.3
1984	2.0	3.1	6.3	13.4	5.3	6.4
1985	1.3	3.1	6.2	11.5	4.2	−0.8
1986	0.8	3.3	6.5	11.4	3.9	−3.5

Year	GDP 1975 = 100	Stockbuilding £M 1975	Manufacturing index 1975 = 100	Current balance 1975 £bn
1979	110.3	1504	104.6	−.6
1980	107.2	−1978	94.8	1.6
1981	104.6	−1713	89.2	1.5
1982	105.4	312	90.2	1.6
1983	106.7	546	90.7	2.9
1984	108.9	860	91.8	3.2
1985	110.3	667	93.0	4.1
1986	111.2	537	94.0	4.9

Simulation 5

This simulation deals with a reduction of direct taxation; it is very similar to the change in indirect taxation with the difference that the strong counter-inflation effects are missing. Higher domestic prices in

Simulation 5: A £5 billion Reduction in Direct Taxation

Year	Real GDP % change	Unemploy- ment millions 4th quarter	Retail prices % change	PSBR		£M3 % change
				£Bn	%GDP	
1979	2.1	1.3	13.4	12.6	7.4	12.7
1980	−2.8	1.9	17.9	12.3	6.4	18.6
1981	−2.7	2.8	11.6	16.0	7.8	11.8
1982	1.0	2.9	9.8	14.2	6.2	10.2
1983	1.5	3.0	8.0	14.2	5.7	10.4
1984	2.0	3.1	8.1	12.8	4.7	10.3
1985	1.5	3.1	8.2	10.3	3.5	9.0
1986	0.9	3.3	8.4	9.0	2.8	8.7

Year	GDP index 1975 = 100	Stockbuilding £M 1975	Manufacturing index 1975 = 100	Current balance £Bn 1975
1979	110.3	1504	104.6	−.6
1980	107.2	−1978	94.8	1.6
1981	104.3	−1753	89.1	1.5
1982	105.3	333	90.2	1.3
1983	106.9	648	91.0	2.3
1984	109.1	1003	92.3	2.4
1985	110.7	766	93.4	3.1
1986	111.7	591	94.5	3.6

Simulation 5 (relative to 4) lead to a smaller, but still substantial, surplus on the balance of payments. The PSBR is slightly lower but both output and unemployment are virtually identical to the levels in Simulation 4.

There are two main lessons to be drawn from the above five simulations. The first and most obvious is that an injection of demand of the order of £5 bn is not going to stop unemployment from rising, no matter how it is spent. The most that can be expected is that unemployment will not rise quite so fast as it would without any easement of policy. The second point is that while increased expenditure has a slightly stronger effect on output and employment than taxation changes, these differences are fairly marginal.

The various components of the simulations can be presented with differing degrees of confidence. The projections for output and employment can be accorded a fairly high degree of confidence while the price effects are subject to a greater degree of uncertainty. In none of the simulations are there large departures from the base case inflation rate (except the reduction in expenditure tax which reduces inflation). This is contrary to the widespread belief that an increase in demand will lead to an increase in inflation. This is not, however, to say that the simulations are wrong. Unemployment in excess of two million may well mean that a slight increase in demand would have no consequences for inflation. There is much uncertainty in this area.

Before proceeding to further simulations, it is important to understand a little about the nature of the model being used. The National Institute model is a large model which has been constructed primarily for forecasting purposes. Although the model is non-linear in many of its equations, its overall performance in simulation is very nearly linear. This means that if a given change in policy has a certain effect then twice the change will have twice the effect, and ten times the change will have ten times the effect. In carrying out simulations it is quite possible within the terms of the model to make huge and rapid changes in policy which could drastically reduce unemployment. All that would be necessary would be a very large injection of demand. It would not, however, be feasible to apply such policies in the real economy. As output rose very rapidly many bottlenecks and supply problems would undoubtedly arise and the shock to confidence at home and abroad of such a huge and sudden change in policy would be severe. These factors would not be captured by the model, so they must be allowed for informally in the type of result which we are seeking to produce. A sudden sharp reduction of unemployment is not a realistic aim, but it does seem reasonable to hope that unemployment could peak in the early part of the simulation period and then begin to decline. Such a path for unemployment and consequently output should not place any extreme strain on the supply side of the economy. Any reflation is bound to have consequences for the PSBR; this has already been shown by the previous simulations. While it has been argued that a close adherence to current PSBR targets is not desirable, there are limits to how large the PSBR can be allowed to grow. Fortunately the rapid fall in the PSBR in the base case does allow considerable room for expansion. It was shown in the last chapter that there is a close association between the fall in the PSBR and the rise in the government's tax revenues from North Sea oil. If a policy of maintaining a constant nominal PSBR were to be pursued, this would be equivalent to channelling these oil revenues into increased expenditure rather than a reduction of borrowing.

A number of simulations were carried out in order to find the kind of policy which would achieve this objective. Simulation 6 presents details of a reflationary package which goes much of the way towards reaching this goal. The simulation consists of a reflation through increased public investment; the amount of the reflation in 1981 is £8 bn, this rises by £6 bn in 1982, and £2 bn in each of the four subsequent years (1981 prices). It may be argued that it would not be possible to increase investment by such a large amount at relatively short notice. This is probably true but any shortfall in actual investment could be made up by general consumption expenditure or tax cuts. The previous simulations show that the effect would be much the same.

Simulation 6

In this simulation, from 1982 output rises at a reasonably acceptable rate. Unemployment actually peaks in 1981 and then falls progressively over the period to produce a figure for 1986 which is more than 1 million below the base case. Considering the size of the reflation the PSBR shows only a modest increase thanks to higher output and employment. Expressed as a percentage of GDP the PSBR is lower in 1986 than either 1979 or 1980. The balance of payments is in deficit from 1982 onwards. This illustrates the problem that any substantial increase in the level of economic activity will quickly erode the balance of payments surplus. However, the deficit makes a devaluation in sterling practical and reasonable. The final simulation will consider the reflation presented in Simulation 6 combined with an 8 per cent devaluation in 1981.

Simulation 7

The small devaluation in 1981 produces a small increase in output and employment over the period, relative to Simulation 6, coupled with an overall improvement on the balance of payments and a modest increase in the rate of inflation. The picture presented here is far from that of an ideal world. Unemployment is still above 2 million in 1986 and output is slightly below its long-term growth path. Nonetheless it does represent a considerable improvement over the base case. Unemployment is falling rather than rising and output is growing at an acceptable rate. The increase in inflation is modest and probably acceptable, and if part of the reflation were in the form of a reduction

Simulation 6: A High and Rising Level of Government Investment

Year	Real GDP % change	Unemploy- ment millions 4th quarter	Retail prices % change	PSBR £bn	PSBR %GDP	£M3 % change
1979	2.1	1.3	13.4	12.6	7.4	12.7
1980	−2.8	1.9	17.9	12.3	6.4	18.6
1981	−0.8	2.5	12.0	17.4	8.2	15.8
1982	2.7	2.4	11.0	17.7	7.3	15.9
1983	1.5	2.4	8.8	18.3	6.9	16.0
1984	2.5	2.3	8.5	18.4	6.3	15.6
1985	2.4	2.3	8.5	18.6	5.7	15.7
1986	1.3	2.3	8.7	19.0	5.3	15.8

Year	GDP index 1975 = 100	Stockbuilding £M 1975	Manufacturing index 1975 = 100	Current balance £Bn 1975
1979	110.3	1504	104.6	−.6
1980	107.2	−1978	94.8	1.6
1981	106.3	−1496	90.9	0.4
1982	109.2	586	93.3	−0.8
1983	110.9	781	94.5	−0.1
1984	113.6	1093	96.3	−0.5
1985	116.4	914	98.2	−0.5
1986	117.9	740	99.8	−0.5

in VAT some of this could be avoided. The question which must be faced, however, is whether or not a reflation, on the scale proposed here, can really be carried through.

In considering this point it is useful first to look at past trends in public investment. Table 8.1 shows the trend in public investment through the 1970s. The reflation in Simulations 6 and 7 was applied in the form of public investment for two reasons. Firstly, the actual levels of public investment fell dramatically through the 1970s and there is a case for restoring these investment levels. It would be a means of carrying forward the benefits of North Sea oil to a later period. Secondly, investment was shown in the earlier simulations to be as good as, or better than, any other form of reflation in terms of reducing unemployment. Table 8.1 shows the trend in public investment through the 1970s.

Simulation 7: A Rising Level of Government Investment plus an 8 per cent Devaluation

Year	Real GDP % change	Unemploy- ment millions 4th quarter	Retail prices % change	PSBR £bn	%GDP	£M3 % change
1979	2.1	1.3	13.4	12.6	7.4	12.7
1980	−2.8	1.9	17.9	12.3	6.4	18.6
1981	−0.5	2.5	12.9	17.3	8.0	16.5
1982	2.6	2.4	12.3	17.1	6.9	15.8
1983	1.5	2.4	9.7	17.3	6.3	16.2
1984	2.5	2.3	9.1	16.9	5.6	15.7
1985	2.3	2.2	8.7	16.6	4.9	16.5
1986	1.3	2.2	8.6	16.8	4.5	16.7

Year	GDP index 1975 = 100	Stockbuilding £M 1975	Manufacturing index 1975 = 100	Current balance £Bn 1975
1979	110.3	1504	104.6	−0.6
1980	107.2	−1978	94.8	1.6
1981	106.7	−1487	90.9	0.2
1982	109.4	533	93.2	−0.6
1983	111.1	741	94.5	0.1
1984	113.9	1059	96.3	−0.3
1985	116.5	896	98.3	−0.4
1986	118.0	744	99.9	−0.5

In the first column of Table 8.1 the figures for public investment are presented from 1970 to 1980 (in 1975 prices). The second column continues this run of figures using the public investment figures of the base simulation. These figures remain constant at the 1980 level. The third column starts in 1970, with the actual public investment figure and then projects a growth path on the assumption that investment grew at the rate of 1.4 per cent per annum, the average growth rate of GDP between 1970 and 1980. The final column is the path of public investment which has been used to generate the reflation in Simulations 6 and 7. In 1975 prices the reflation is £4 bn in 1981, £7 bn in 1982, £8 bn in 1983, £9 bn in 1984, £10 bn in 1985, and £11 bn in 1986. The largest single change in investment levels comes in 1981 when total public investment is raised to £10.5 bn. This is almost

Table 8.1: Public Investment Figures

£Bn 1975	Actual public investment	Base simulation public investment	1.4% growth in public investment from 1970	Simulation public investment
1970	8.8		8.8	
1971	8.6		8.9	
1972	8.0		9.0	
1973	8.9		9.2	
1974	9.0		9.3	
1975	8.9		9.4	
1976	8.8		9.6	
1977	7.7		9.7	
1978	7.0		9.8	
1979	6.7		10.0	
1980	6.4		10.1	
1981		6.5	10.2	10.5
1982		6.5	10.4	13.5
1983		6.5	10.5	14.5
1984		6.5	10.7	15.5
1985		6.5	10.8	16.5
1986		6.5	11.1	17.5

exactly the level which investment would have reached on its steady growth path. The subsequent increases in investment do take it well above its trend values but even these figures can be justified on the grounds of carrying out past public investment which did not take place. Total investment from 1970-86, on the 1.4 per cent growth path, would have been £167 bn. The total of actual investment from 1970-81 plus the Simulation 7 investment plan would be £174 bn. This reflation package represents an increase in total investment of only £7 bn in the 1970-86 period over a steady growth trend. There may be many administrative difficulties arising from the sharp changes in investment levels proposed in the reflationary path. This may well mean that part of the reflation should work through some other instrument, tax cuts or increases in government consumption expenditure. This need not greatly affect the overall impact of the reflation.

The conclusion which emerges from Table 8.1 is that the actual size of the proposed reflation is not, in fact, unreasonable given the recent decline in public investment. The other aspect of Simulation 7 which might cause concern to some people is the size of the PSBR. Although the PSBR does fall over the period when it is expressed as a

Table 8.2: The Relationship between GDP and the Cumulated PSBR

Year	Actual figure to 1980	Base simulation	Simulation 9
1965	1.05		
1970	0.84		
1975	0.64		
1978	0.58		
1979	0.58		
1980	0.56		
1981		0.58	0.58
1982		0.58	0.58
1983		0.57	0.58
1984		0.54	0.58
1985		0.51	0.58
1986		0.48	0.57

percentage of GDP, its absolute value in nominal terms remains quite high. In Chapter 6 it was argued that the nominal value of the PSBR was not a crucial factor, but that the relationship between the national debt (or the cumulated PSBR) and GDP was the important indicator of the stance of government policy. Table 8.2 gives a brief summary of how this ratio has moved recently and indicates the path which the ratio would take in the base simulation and in Simulation 7.

The ratio of the cumulated PSBR to GDP has been falling ever since the war. The base simulation shows that this decline would continue under unchanged policies. This illustrates the deflationary nature of present policies. The simulated reflation merely maintains the ratio at an unchanged level. The size of the cumulated PSBR does not grow in real terms even under such an apparently expansionary policy. Viewing the PSBR figures in this way, the Simulation 7 proposals do not seem either excessive or unrealistic.

The oil production levels and tax revenues which lie behind the simulation are identical to those of the base case (outlined in Table 7.10, Chapter 7). The way these revenues are applied is, however, radically different. In the simulations the GDP-national debt ratio is being broadly held constant. This means that the government oil tax revenues are no longer being directed towards a reduction in the PSBR but are rather being directed towards increased levels of direct investment. This increase in government investment leads to a general increase in economic activity which in turn wipes out the large current account surplus in the base case. This represents a switch in investment from foreign to domestic investment. The oil revenues are being turned towards domestic investment in this alternative policy; this is producing large multiplier effects and reversing the direction of the economy and bringing the recession to an end.

The oil benefit plays an important part in this alternative development. In the absence of oil the current account deficit in 1986 would stand at nearly £11 bn (1975 prices) and the PSBR would be around £35 bn (current prices). Neither of these would be acceptable. A massive devaluation and general deflation would almost certainly be the government's response to such a situation. While oil is not playing a superficially obvious part in the strategy outlined in Simulation 7, without oil the alternative strategy would be far more difficult to carry out.

Reflation and Inflation

The objection to any suggestion of reflation is that it will inevitably lead to increased inflation. This may be so, but the amount of the increase is very uncertain, and the cost of any additional inflation may be far less than the cost of long-term and increasing unemployment.

At a theoretical level there is a sharp division of inflation theories into two groups. The first group links inflation to some aspect of the real economy, such as unemployment, and the second takes the extreme monetarist view that inflation is determined only by changes in the money supply. The monetarist view would, therefore, involve no direct link between fiscal reflation and an increase in the rate of inflation. If the money supply were held down during the reflation then there would be no reason for the rate of inflation to rise. The argument typically presented by supporters of this approach is that a reflation would necessarily increase the PSBR and that this would in turn produce an increase in the money supply. This mechanism was discussed in Chapter 6, where we found that it lacked both theoretical and empirical support. An increase in the PSBR will not necessarily lead to an increase in the money supply so even if the argument for a strong link between money and inflation were valid there would not be a necessary link between fiscal reflation and an increase in inflation.

In the more traditional view the rate of inflation responds to some indicator of total demand in the economy. The most common view, expressed in the Phillips curve, relates the rate of inflation to expected inflation and the level of unemployment. If unemployment is above some critical level of unemployment then inflation will fall, if it is below this level then the rate of inflation will rise. On the basis of this theory as long as unemployment is above the critical rate inflation will continue to fall even if the level of unemployment is being reduced by reflation. The crucial factor in this theory is, of course, how much unemployment constitutes the critical level. During the 1950s and 1960s this would have been estimated at something like 2 per cent. Some commentators suggest that during the 1970s increased levels of unemployment benefits and changes in trade union organisation have caused a rise in the natural rate. This is almost an impossible argument to refute yet we find it highly implausible that the 'natural' rate of unemployment can have risen to anything like the current level. So, as long as we are prepared to assume that the 'natural' rate is significantly below current levels of unemployment (around 12 per cent), there is no contradiction between reflation and a reduction in inflation on the basis of this theory.

There is, however, an offshoot of the basic Phillips curve analysis which would contradict this. During much of the 1970s researchers undertaking empirical work have failed to find a well-defined conventional Phillips curve but a number have found a relationship between the rate of inflation and *changes* in unemployment. So that when unemployment rises it is found to reduce the rate of inflation and when it falls the rate of inflation rises. According to this model of the inflation process any reflation which reduces unemployment must increase the rate of inflation. On the basis of this theory there is no way to reduce unemployment without increasing inflation. Either we must permanently tolerate the current levels of unemployment or we must eventually suffer the increase in inflation which will occur when unemployment falls. If we intend ultimately to reduce unemployment then it is better to do it now rather than later because we will have to suffer the same cost in terms of increased inflation now or later, but by doing it now we escape the cost of the unemployment itself.

Even from this brief description of the theories underlying the determination of inflation it is apparent that there is a great deal of uncertainty in this area. Before leaving this topic it is worth briefly considering the actual evidence for the UK during the 1970s. Figure 8.1 shows the relationship between inflation and unemployment during the period 1970-82. The most important point made in this figure is that we really have very little evidence about the inflationary consequences of a fall in unemployment. Unemployment actually

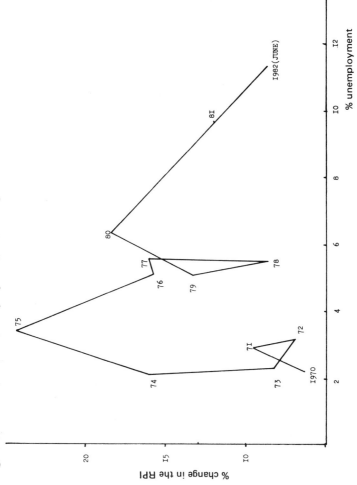

Figure 8.1: Inflation and Unemployment in the UK, 1970-1982

fell in only four years: 1973, 1974, 1978 and 1979. But of these four years three are highly distorted by oil price shocks and the government's own policies in 1973, 1974 and 1979. Even 1978, which shows a small fall in unemployment coupled with a large fall in inflation, is subject to a considerable degree of distortion caused by the contemporary prices and incomes policy. This graph does, however, suggest that, with the exception of the 1979/80 period (which was distorted by the increases in oil prices and VAT), there has been a tendency for the rate of inflation to fall since 1975. In any case with unemployment of the order of 10 per cent, an increase in inflation is likely to be fairly limited.

A Long-term Perspective

So far we have considered medium-term policy options, i.e. the next five years, but North Sea oil is likely to be a significant factor in the UK economy for the next 40-50 years. How far should current economic policy be affected by this long-term prospect of oil wealth? Should we be taking positive steps to prepare for the end of North Sea oil in terms of investing in alternative energy sources or by building up the industrial sector of the economy?

Economic theory has traditionally analysed the long-term development of an economy in a very different way from the short-term modelling methods which have been used for economic forecasting and policy formation. Long-term analysis has generally focused on the factors affecting the growth and development of an economy on the underlying assumption that the economy is at full employment. The determinants of growth are then found to be such factors as technical progress, population growth and the percentage of output devoted to investment. Population growth is of negligible importance here as it is largely outside the control of government. The other two factors are both crucial to the growth process and are difficult to control. The normal determinants of both technical progress and investment are generally taken to be interest rates and industrial profitability, both of which are closely linked to the level of aggregate output and short-term economic policy. A government which adopts a policy of high interest rates and deflationary budgets is providing an economic environment which is hostile to both investment and technical progress. The short-term policies outlined earlier in this chapter would tend to raise investment levels as aggregate demand increases and the loosening of tight monetary control would allow scope for the reduction of interest rates. The message here would seem to be that if we look after the short-term prospects for the economy the long term will look after itself.

What about North Sea oil? The crucial question here is how long the oil is going to last. If North Sea oil may be regarded as permanent then we do not need to make any special provision for its end. But if it is assumed to be only a very temporary thing then it would probably be desirable to prevent the automatic effects outlined in Part II of this book from taking place. So if oil were expected to run out at the end of this decade we might certainly adopt a policy of encouraging foreign investments. Then the exchange rate would be held down and domestic industry protected from any relative decline, and when the oil was gone industry would remain strong. If the oil sector is regarded as permanent this problem does not arise, and indeed the structural adjustments which would occur in the non-oil sector of the economy may actually be seen as desirable. The analysis of the first part of this book makes it clear that oil will be a major factor in the UK economy for a very long period. We feel that an industry which can confidently expect 40 or 50 years of reasonably high production levels must be regarded as semi-permanent, so to take special measures to protect the non-oil sectors of the economy would be undesirable.

This is still not the whole story. Oil, both in a national and an international context, is becoming a scarce commodity; its relative scarcity is already evident from the large increases in its price in the late seventies. In a perfect market economy this situation would not call for any special response on the part of the government; the high oil price would cause an increase in research into alternative energy sources. Subsequently as new forms of energy technology became available we would expect industry quickly to develop these techniques. In the real world, however, a large part of the pure research which would set off this process is funded either directly by the government or by grants from various large foundations. Neither of these is likely to respond to the market's price signals. Government funding for such research is determined mainly by the overall stance of its fiscal policy. If it is aiming to cut expenditure then such pure research projects are a relatively easy area in which to make reductions. In 1981/2 the Department of Energy spent £45 million on non-nuclear energy research; it is proposed to cut this figure to £42 million in 1982/3 and then hold it constant in nominal cash terms. When these figures are compared with the £620 million spent by the Department of Industry supporting British Leyland in 1981/2, the relatively low priority which energy research is given becomes obvious. Of course, if the fiscal stance of the government were to be reversed then it would be precisely such areas as energy research which would benefit. The areas in which it is easy to make cuts tend also to be the easiest and most profitable to expand, so it would seem that the short-term policies outlined earlier would have beneficial long-term effects.

Conclusion

During 1981 the main calls for reflation suggested an injection of about £5 bn in 1981 prices. This chapter has considered the likely effect of such an injection over a medium-term horizon. The conclusion which emerges is that the actual method of reflation is of relatively minor significance and that a reflation of £5 bn cannot be expected to halt the rise in unemployment or return output to its trend growth path. This conclusion led to a search for a reflationary package which would reverse the trend in unemployment and restore output to a more normal rate of growth. It was found that a reflation which restored public investment to its trend levels and maintained the cumulated PSBR-GDP ratio was sufficient to bring about this effect.

9 CONCLUSIONS

It may be convenient to the reader to have the conclusions of this study brought together in a short final chapter. We have concentrated on three aspects of the subject of North Sea oil and gas and their relationship to the economy as a whole. First we considered our oil and gas resources as a problem of resource management. We then looked at the effect of the development of these resources on the industrial sector of the economy. Finally we analysed the links between oil and general economic policy.

The Management of Oil and Gas Resources

Many varying estimates have been made of the total remaining reserves of oil and gas on the UK continental shelf. Our conclusion in this most uncertain field is that there is probably something approaching 4,000 million tonnes of oil and 1,600 million tonnes of gas in terms of oil equivalent. Reserves of these amounts would last for nearly 60 years in the case of oil and about 40 years in the case of gas at present rates of consumption.

One of the main issues arising from these great discoveries is that of depletion policy. A rational depletion policy requires decisions on how much of the benefits should be consumed currently and how much saved for the future. It also requires a further decision on how best to invest any savings which are to be made.

Our conclusion on the first point is that at least 1 to 2 per cent of the stock can legitimately be consumed every year, since this is only the income from the capital asset, which should appreciate by that much on average year by year. Our conclusion on the second point is that because of the insecurity of world oil prices, one of the best investments is probably oil in the ground. Combining these two considerations with the practical aspects of development we conclude that depletion policy should be aimed at securing a level of production over the next few years not much in excess of the rate of domestic oil consumption.

Oil and the Manufacturing Sector

There has been considerable debate in recent years over the effect of oil on the manufacturing sector. Does the growth of oil production necessarily lead to an absolute or relative decline in manufacturing output? Has oil caused the fall in British manufacturing output since 1979? We answer these questions in the following way.

Firstly, on simplified assumptions, it is generally the case that the development of a natural resource such as oil, which displaces imports and is itself exported, will reduce the market for a country's output of traded goods and services, typically manufactures. However, in the particular case of the UK the improvement in the balance of payments due to oil was superimposed on a sizeable deficit which resulted in part from the rise in world oil prices in 1973/4. Thus the effect of oil in the case of the UK was rather to eliminate the deficit than to create a surplus in the balance of payments.

There were no forces making for a continuing decline in manufacturing output as a result of oil production: the rise in oil production merely obviated the need to restore the balance of payments by building up exports.

By the early 1980s the balance of payments was in large surplus, but this owed much to the recession and the low level of imports. The sharp decline in manufacturing output in 1980/82 was mainly a result of deliberately restrictive economic policies. The high exchange rate played some part in the decline of manufacturing output and oil production contributed to the high exchange rate. However, it is very difficult to disentangle the effect of oil on the exchange rate from all the other factors at work, especially high interest rates.

Without the oil no doubt we should have needed to sell more non-oil exports, i.e. manufactures. But the level of manufacturing output in the absence of oil would have depended on how the government of the day handled the massive balance of payments problem from which we would have been suffering. Such an exercise in imagining what might have been could not be conclusive since it would depend on arbitrary assumptions about economic management in the totally different circumstances of UK without oil.

It may be that oil did have some depressing effect in the early 1980s on the manufacturing sector in the UK, though it was probably only a minor factor. However, such depressing effects as oil production had could have been avoided if the government had followed a more expansionary economic policy. This would have meant the

the disappearance of the surplus in the balance of payments and a lower exchange rate, hence more demand for British manufactures, both at home because of the higher level of activity and from abroad because of the more competitively priced exports.

Oil and Economic Policy

Many people had expected oil to bring great benefits to the economy: in the event the years of self-sufficiency have been years of prolonged recession and growing unemployment. Our explanation for this paradox is that there was a coincidence between the growth of oil production to high levels in recent years and the adoption by the government of deflationary policies from 1979 onwards. Oil may have contributed to the higher exchange rate of the early 1980s, which tended to depress the economy; however, the main cause of the higher exchange rate was the monetary policy which kept interest rates high. Another factor in raising the exchange rate was the surplus in the balance of payments on current account, which resulted from the low level of activity at home.

Our conclusion is that oil played only a minor part in the long recession beginning in 1980 and its effects could easily have been offset. The government did not try to offset these effects: the high exchange rate fitted well into the government's economic strategy, which had the reduction of inflation as its overriding aim.

The government did not deliberately seek to depress the economy. Ministers believed that control of the money supply and reduction of the budget deficit (with the help of growing oil revenues) would bring down inflation with but a temporary check to economic activity.

The deep recession has made plain that this belief was mistaken. Even before the recession the monetarist ideas which inspired economic policy were criticised by mainstream economists who foresaw that such policies would cause stagnation and massive unemployment.

As the policies of monetary and fiscal restraint were steadily applied year after year in accordance with the government's medium-term financial strategy, more and more proposals were made for alternative policies to restore growth and prosperity. Most of these proposals were fairly modest and amounted to a limited reflation of demand. We used the National Institute model of the economy to simulate the effects of several different reflationary measures, each worth £5 bn, ranging from increases in government expenditure to reductions in various kinds of taxes.

The results indicate that a stimulus of this order would be insufficient to move the economy on to a satisfactory growth path. Unemployment would still go on rising whichever of the various measures were adopted. To achieve a fall in unemployment a much bigger stimulus would be needed. We suggest a substantial and rising increment to public investment coupled with a small devaluation. On this basis it should be possible to keep the rate of growth up to 2½ per cent and bring unemployment down slightly.

With this policy inflation would probably increase somewhat above its current rate instead of declining further as it would on the government's present policies. Since unemployment would remain very high there should not be any great increase in wage pressures: but there would be some, and the lower exchange rate would also contribute to somewhat higher prices.

The Effects of Oil on the Economy

The book has been divided into three sections but the links between them are of great importance. We argued in the final section that economic policy should not be directly linked to the North Sea sector but that it should be determined along macroeconomic lines within the resource constraints which are imposed on the economy. North Sea oil affects economic policy by changing these constraints; this was discussed in detail in the first part of the book.

The middle section examined the effects on industry of developing the oil sector but it left out the response of the government, which makes up the final part.

Given these complications a simple summary of the effects of oil is meaningless. If we ask 'Has oil been a good thing?', we may answer 'Yes', in terms of its direct contributions to GDP and the balance of payments, or 'No' in terms of the automatic exchange rate effects on industry and employment. The answer depends on how we view the government's economic policies during the period of oil build-up. A more appropriate question might be 'Has the government managed the economy correctly during the build-up of North Sea oil?' To this question we answer 'No' and we go on to propose an alternative policy which would make better use of the North Sea endowment.

REFERENCES

Bank of England (1980) 'The North Sea and the United Kingdom Economy, Some Long-Term Perspectives and Implications', Bank of England Quarterly *Bulletin*, Vol 20, No 4

Bank of England (1982) 'North Sea Oil and Gas: A Challenge for the Future', Bank of England Quarterly *Bulletin*, Vol 22

Banks, F.E. (1980) *The Political Economy of Oil*, Lexington

Barker, T.S. and Brailovstiy, V. (1981) *Energy, Industrialistion and Economic Policy*, Academic Press, London

Barouch E. and Kaufman, G.M. (1976) 'Oil and Gas Discovery Modelled as Sampling Proportional to Random Size', Alfred P. Sloan School of Management Working Paper, MIT

Barouch, E. and Kaufman, G.M. (1976) *Probabilistic Modelling of Oil and Gas Discovery In Energy: Mathematics and Models*, Siam Publishing

Blackaby, F.T. (1979) *De-Industrialisation*, Heinemann

Brooks, S. (1981) The Economic Implications of North Sea Oil, NIESR Discussion Paper No 38

Byatt, I., Hartley, N., Lomax, R., Powell, S. and Spencer, P. (1982) North Sea Oil and Structural Adjustment, Treasury Working Paper No 22

Cook, P.L. and Surrey, A.J. (1977) *Energy Policy: Strategies for Uncertainty*, Martin Robertson

Cook, S.T. and Jackson, P.M. (1979) *Current Issues in Fiscal Policy*, Martin Robertson

Corden, W.M. (1981) 'The Exchange Rate, Monetary Policy and North Sea Oil: The Economic Theory of the Squeeze on Tradeables', Oxford Economic Papers, Vol 33

Dafter, R. and Davidson, I. (1981) North Sea Oil and Gas and British Foreign Policy, Chatham House Papers No 10

Dasgupta, P.S. and Heal, G.M. (1979) *Economic Theory and Exhaustible Resources*, James Nisbet and Co and Cambridge University Press

Davidson, P. (1979) 'Oil Conservation: Theory vs. Policy', *Journal of Post-Keynesian Economics*

Department of Energy (1978) Energy Policy: A Consultative Document, HMSO, Cmnd. 7101

Department of Energy (1982) *Development of Oil and Gas Resources of the United Kingdom*, HMSO

Eastwood, R.K. and Venables, A.J. (1982) 'Macroeconomic Implications of a Resource Discovery in an Open Economy', *Economic Journal*

Forsyth, P.J. and Kay, J.A. (1980) 'The Economic Implications of North Sea Oil Revenues', *Fiscal Studies*

Forsyth, P.J. and Kay, J.A. (1981) 'Oil Revenues and Manufacturing Output', *Fiscal Studies*

Gray, L.C. (1914) 'Rent under the Assumption of Exhaustibility', *Quarterly Journal of Economics*

Gregory, R.G. (1976) 'Some Implications of the Growth of the Mineral Sector', *Australian Journal of Agricultural Economics*

Hamilton, A. (1978) *North Sea Impact: Offshore Oil and the British Economy*, International Institute for Economic Research

Hotelling, H. (1931) 'The Economics of Exhaustible Resources', *Journal of Political Economy*

Houthakker, H.S. (1980) 'The Use and Management of North Sea Oil', in Caves, R.E. and Krause, L.B. (eds.), *Britain's Economic Performance*, Brookings Institute, Washington

Johnson, C. (1977) 'North Sea Energy Wealth 1965-1985', *Financial Times*

Johnson, P.S. (1980) *The Structure of British Industry*, Granada

Kemp, A.G. and Cohen, D. (1980) 'The New System of Petroleum Revenue Tax', *Fiscal Studies*

Kubbah, A.A. (1974) *OPEC Past and Present*, Petro-Economic Research Centre

Meade, J.E. and Russell, E.A. (1957) 'Wage Rates, the Cost of Living and the Balance of Payments', *Economic Record*, Vol 33

Motamen, H., Hall, S. and Strange, R. (1983) 'The macroeconomics of North Sea Oil in the UK economy', Heinemann

Motamen, H. and Strange, R. (1981) *UK Oil Revenue: The medium-term outlook*, Energy Policy

Nordhaus, W.D. (1979) *The Efficient use of Energy Resources*, Yale University Press

Odell, P.R. and Rosing, K.E. (1975) *The North Sea Oil Province: an Attempt to Simulate its Development and Exploitation 1969-2029*, Kogan Page

Odell, P.R. and Rosing, K.E. (1980) *The Future of Oil: a Simulation*

Study of the Inter-Relationship of Resources, Reserves and Use, 1980-2080, Kogan Page

Richardson, R. (1980) The North Sea and the United Kingdom Economy: Some Long-Term Perspectives and Implications, The Ashridge Lecture, November

Robinson, C. and Morgan, J. (1978) *North Sea Oil in the Future*, Macmillan

Scott, M.Fg. (1982) The Impact of Changes in the Real Price of Fuels on Incomes in the United Kingdom, unpublished, Nuffield College, Oxford

Solow, R.M. (1974) 'The Economics of Resources or the Resources of Economics', *American Economic Review*

Whiteman, J. (1980) *North Sea Oil and the U.K. Economy*, NEDO Economic Working Paper, No 5

Worswick, G.D.N. (1980), 'North Sea Oil and the Decline of Manufacturing', *National Institute Economic Review*, November

INDEX